PHYSICS DEPT:

ASTRONOMY FOR GCSE

COLLYER'S
SIXTH-FORM COLLEGE

Astronomy for GCSE

NEW EDITION

Patrick Moore
and
Chris Lintott

Duckworth

This edition first published in 2001 by
Gerald Duckworth & Co. Ltd.
61 Frith Street, London W1D 3JL
Tel: 020 7434 4242
Fax: 020 7434 4420
Email: enquiries@duckworth-publishers.co.uk
www.ducknet.co.uk

A catalogue record for this book is available from the British Library

ISBN 0 7156 2969 7

Typeset by Derek Doyle & Associates, Liverpool
Printed in Great Britain by
Ebenezer Baylis & Son Ltd, Worcester

Contents

Foreword

by Dr Frank Flynn

'Why GCSE Astronomy?' 'Haven't we already enough subjects in the overcrowded school curriculum?' 'What relevance has it anyway to the practical issues of earning a living in a modern High-Tech society?'

What a pity when people think of the study of the universe in these utilitarian terms. It is true that astronomy has tended to be treated as a Cinderella subject in schools, relegated to the twilight fringes of the curriculum, and to club activities after school. But this is very largely due to the primary importance of studying the fundamental sciences in core time. It is good to see that astronomical material has been consciously built into the National Curriculum at all levels, and there is some impressive work going on in primary schools as well as secondary.

Children of all ages have enquiring minds, and I have always found it remarkable how genuinely interested and well informed even quite young children can be. Children have a habit of showing up adults where astronomy is concerned. Astronomy differs from all other disciplines in that it takes us on a journey right away from our own home planet, deep into the universe, and we cannot fail to marvel at the immensity and wonder of it all and the tininess of Planet Earth.

The GCSE Astronomy syllabus and examination is administered by EDEXCEL, the former University of London Examinations and Assessments Council. Recently major revisions have been made to update the syllabus, and to cater for a wide range of students. These include those at school who may be following an astronomy course as a science or general studies option in the 10th or 11th years of secondary schooling, sixth-formers managing to fit in another subject to complement their A level course and, not least, adult students who may be studying astronomy as a leisure activity, attending an astronomical society or evening class, or simply working on their own.

Some may wonder if an examination tends to kill the enjoyment of such an activity. I am sure this is not so. The whole concept of the course is entirely friendly. The goal of the GCSE certificate stimulates the student to read, tackle interesting practical coursework, for which there is a wide range of projects, and meet and discuss with kindred spirits.

Many students follow the course and take the examination for personal satisfaction, rather than to obtain a specific qualification, but for those who need GCSE results, a grade C in this subject is equivalent to a grade C in any of the other sciences.

There are many excellent books on the market on every aspect of astronomy, but few which draw together the main strands of a GCSE level course in a convenient and readable way. For this reason this revised edition by Patrick Moore and Chris Lintott is extremely welcome. It is written in a personal and chatty style, with something for everybody, and as usual a great deal of encouragement. There are review questions and suggestions for practical work at the end of each chapter, and the glossary of technical terms at the end is particularly handy. I commend this useful book to you, and wish you much enjoyment and satisfaction in following the course.

Frank Flynn
Formerly Chief Examiner for GCSE Astronomy

Acknowledgements

Obviously, the production of a book such as this leaves us with many people to whom we are profoundly grateful. There is only space to mention a few of them here but nevertheless we thank all of them. For the first edition, special thanks were due to Paul Doherty for the excellent illustrations, many of which are used once more here, Iain Nicolson, Dr Ron Maddison and Dr John Mason.

Deborah Blake from Duckworth deserves thanks for tireless support, and the book also would not have been possible without assistance in a whole variety of ways from Adam Corrie and Catherine Burgess. Finally, we thank Dr Frank Flynn for his Foreword and endorsement of this book.

Cambridge, October 2000

P.M.
C.J.L.

ix

Units of Measurement

At present (2000) Britain is going through a transitional period. The familiar Imperial units are still in use, but the official policy is to replace them with Metric. We have therefore used Metric in this book, even though this sometimes involves some 'rounding off'. For example, many people own 3-inch refracting telescopes. In Metric, this is equivalent to 7.62 centimetres, which we have usually given as 7.6.

Conversion factors

To convert	Multiply by
kilometres to miles:	0.621
metres to feet:	3.281
centimetres to inches:	0.394
millimetres to inches:	0.039
square km to square miles:	0.386
square metres to square feet:	10.764
square mm to square inches:	0.002
tonnes to tons:	0.984
kilograms to pounds:	2.205
grams to ounces:	0.035
miles to kilometres:	1.609
feet to metres:	0.3048
inches to centimetres:	2.54
inches to millimetres:	25.4
square miles to square km:	2.60
square feet to square metres:	0.093
square inches to square mm:	645.16
tons to tonnes:	1.106
ounces to grams:	28.35

The English 'billion' is 'one million million', while the American 'billion' is 'one thousand million'. This can still cause confusion, so we have avoided using 'billion' at all.

Scientifically, temperatures are given in degrees Centigrade or Celsius (°C) and not in Fahrenheit (°F). If you want an approximate conversion, then:

$$F = (C \times 2) + 30$$
$$C = (F - 30) \div 2.$$

The Kelvin scale begins at absolute zero: –273°C. To convert from C to K, add 273 to the Celsius value. Thus O°C = 273K.

1
Introduction to the First Edition

Textbooks are sometimes as dry as dust. I am setting out to write a textbook, but I hope that it will be the reverse of dull – indeed, astronomy is one of the most fascinating of all subjects. The skies are all around us, and there is always something new to see. Moreover, astronomy is fun.

For many years astronomy was an O Level subject in GCE, and as long ago as 1970 it was suggested that I might write a book about it, suited to those who were anxious to take the examination. I was delighted to accept, and I think my book, *Astronomy for O Level*, was widely used – particularly as astronomy is not often taught as a separate school subject, though naturally it must come into any full science syllabus. My original O Level book was revised and reprinted several times, but now that the GCE has been replaced by GCSE it is time for a complete re-write, if only because the astronomy syllabus itself has been altered, giving more emphasis on practical work. What I have tried to do is to provide a text which will be sufficient for any candidate to cover the whole syllabus without going into more detail than is necessary; remember the boy who read a book about penguins and then commented, 'This book told me more about penguins than I really wanted to know'. Do not be put off by the small amount of mathematics involved. There is nothing frightening about it.

I first became interested in astronomy when I was six years old (in 1929!). My first step was to read some books and

learn the basic facts. Next, I borrowed an outline star-map, went outdoors after dark and learned my way around the sky, which is not nearly so difficult as might be thought. Then I obtained a pair of binoculars and spent many weeks looking at the Moon and the stars. I was also lucky in being able to buy a telescope, which meant much less of a financial outlay than it does today. I will have more to say about this later. I still think that I tackled matters in the right way; at the present time I would also recommend joining an astronomical society, which is easy inasmuch as most large cities and towns have associations which are open to everyone.

I am writing for GCSE students, not for would-be professional astronomers. Yet I receive many letters from readers who are anxious to turn professional in due course, and I feel that I should make one important point without delay. The professional astronomer is, above all, a mathematician and a physicist; he spends relatively little time in looking through the eye-end of a telescope, and almost all research is carried out either by photography or, increasingly, by electronic devices. Moreover, the emphasis is upon the distant stars and star-systems rather than on our near neighbours, the Moon and planets.

To anyone who wants to become a professional astronomer, a science degree is absolutely essential. Oddly enough, this degree need not be in pure astronomy. Many professionals begin by taking their original qualification in physics, for example. But to the question 'How can I become a professional astronomer without taking a degree?' I can only answer, in all honesty, 'You can't'. This is an age in which qualifications are all-important. On the other hand, anybody can become a useful amateur astronomer and will derive immense enjoyment from it. Here I can speak with authority, since I have been an amateur all my life and would never consider trying to change my status now. Bear in mind, too, that astronomy is one of the very few sciences in which the amateur can carry out valuable research.

For the moment, then, let us concentrate upon astronomy for GCSE. To take the examination is not difficult; I wish you all success, and I hope that I will be able to help you.

1990   Patrick Moore

Introduction to the Second Edition

Since this book last appeared, in 1990, a great deal has happened. There have been major changes in the GCSE Astronomy syllabus, and, of course, important developments in astronomical science. This means that the last edition is now out of date.

For this new version, the text has been completely overhauled, and largely rewritten, by the two of us working together. We hope that the result is acceptable, and that it will be of use both to those who are anxious to take the GCSE examination and to those who simply want to learn more about this fascinating subject.

2000   Patrick Moore
           Chris Lintott

2
Fundamentals

Before starting out, it may be as well to 'clear the air' with a chapter devoted to basic facts. And first, let us stress that there is absolutely no connection between astronomy and the false science of *astrology*. Astrologers claim that by working out the positions of the Sun, Moon and planets at the time of a person's birth, they can tell his (or her) character and destiny. An astrological chart (a horoscope) may look impressive, but it means nothing at all, and the only polite word to describe astrology is 'rubbish'. There may have been some excuse for it long ago, but there is none today. The best that can be said of it is that it is fairly harmless when treated purely as a parlour game.

The Earth is a *planet*, a world 12,756 km in diameter, moving round the Sun in a path or *orbit* at a distance of almost 150,000,000 km on average. This may sound an enormous distance, but it is not much to an astronomer, who is used to dealing with distances and time-spans quite beyond our everyday experience. The Sun is a *star* – a huge globe of hot gas, big enough to swallow up over a million Earths. It is creating its own energy, and the same is true of the stars visible at night, which are themselves suns. Our Sun, which looks so glorious, is nothing more than an ordinary star. It appears so much more brilliant than the other stars simply because it is much closer to us. Even the nearest star beyond the Sun is approximately 40 million million km away.

(In Imperial measure there used to be a very convenient scale model. The Earth's mean distance from the Sun – 150 million km – is known as the *astronomical unit*. Represent this by one inch, and the nearest star will be just over 4 miles away. In Metric, the corresponding values are 2.5 cm and 6.7 km.)

The Earth is not the only planet in the Sun's family or *Solar System*. There are eight others, moving at different distances at different speeds. Mercury and Venus are closer to the Sun than we are; next come the Earth and Mars, after which there is a wide gap before we come to the four giant planets, Jupiter, Saturn, Uranus and Neptune, together with a curious little world, Pluto, which seems to be in a class of its own. With the naked eye the bright planets look like stars, but they are not nearly so important as they look, and they have no light of their own; they shine only because they reflect the rays of the Sun.

Some of the planets have secondary bodies or *satellites* moving round them. The Earth has one satellite, our familiar Moon. Like the planets, it has no light of its own, and obviously the Sun can illuminate only half of it at any one time, which is why the Moon shows its regular *phases* or apparent changes of shape from new to crescent, half, three-quarters and full. It has a diameter of less than 3500 km, but it is much closer than any other astronomical body, and moves at a mean distance of only 384,000 km from the Earth. Other planets have whole families of satellites – seventeen in the case of Saturn.

The Solar System also includes bodies

of lesser importance. The *asteroids* or minor planets are dwarf worlds, less than 1000 km across, most of which move round the Sun between the orbits of Mars and Jupiter. *Comets* are flimsy, ghostlike bodies made up of icy nuclei together with thin gas and 'dust'; they move round the Sun, but their orbits are generally very eccentric. *Meteors*, of sand-grain size, are cometary débris. They are visible only when they dash into the Earth's upper air and burn away by friction, producing the luminous streaks which we call shooting-stars. Larger bodies can strike the Earth without being burned away, and may produce craters; they are then termed *meteorites*. However, a meteorite is not simply a large meteor, and is more closely related to the asteroids.

The star-patterns or *constellations* have no real significance. The stars are at very different distances from us, and we are dealing with line-of-sight effects. The constellations do not change appreciably even over long periods because the stars are so remote that their individual or *proper* motions are very slight. Their distances are so great that to measure them in kilometres or miles would be clumsy, just as it would be awkward to give the distance between London and Manchester in centimetres. Instead, astronomers use units based on the velocity of light, which is approximately 300,000 km per second (usually written nowadays as 300,000 km sec^{-1}). In a year, light can cover approximately 9.46 million million km, and this is the astronomical *light-year*. The nearest star beyond the Sun is just over 4 light-years away, but most are much further off. For example, the distance of the Pole Star is 680 light-years – so that we now see it not as it is today, but as it used to be 680 years ago, around the time of the Battle of Bannockburn!

The members of the Solar System are much closer, and so they do seem to wander about slowly from one constellation to another, though admittedly they keep to certain well-defined parts of the sky. Because the Earth takes one year to complete one orbit round the Sun, the Sun takes a year to make a full circuit of the sky.

The Sun is a member of the star-system or *Galaxy*, which contains roughly 100,000 million stars of all types – some hotter than the Sun, some cooler; some much more luminous, others much less so. There are also patches of gas and dust called *nebulae* (Latin, 'clouds'), as well as a vast amount of thinly-spread interstellar matter. The Galaxy is a flattened system, with a central bulge. The Sun lies near the main plane; when we look along this plane we see many stars in almost the same line of sight, causing the lovely, shining band which we call the Milky Way.

Our Galaxy is not the only one. Far away, in most cases many millions of light-years from us, we can see others – such as the spiral galaxy in the constellation of Andromeda, over 2 million light-years away. Nowadays we can use telescopes powerful enough to detect systems well over 10,000 million light-years away.

We do not know the full extent of the universe; but we have found that apart from a few of the nearest systems, all the galaxies are moving away from us, so that the whole universe is expanding. How the universe began, and how it will end – if, indeed, it will end at all – are problems that we have yet to solve.

These, then, are the basic facts. Now we are ready to look back in history, and see how this fund of knowledge has been built up.

Questions

1. Which is larger: the Sun or the Moon?
2. What is a constellation – and are the stars in any particular constellation genuinely associated with each other?
3. What is a shooting-star?
4. Define the astronomical unit.
5. If, on a scale model, the distance between the Earth and the Sun is given as 5 cm, how far away must we put the nearest star beyond the Sun?
6. What is a galaxy?
7. How long does the Sun take to move right round the sky?
8. Why does the Moon appear so brilliant?
9. Explain what is meant by a light-year, and give its approximate value.
10. What are the asteroids?

3
Astronomy through the Ages

Astronomy must surely be the oldest science in the world. Even our remote cave-dwelling ancestors must have taken an interest in the sky; many of them believed the Sun and Moon to be gods, while the stars were tiny lights fastened on to the inside of an invisible crystal dome. The Earth, of course, was flat, and lay motionless in the exact centre of the universe.

All this was natural enough. The Sun sends us almost all our light and heat, and makes life possible; the world really does look flat, and the various bodies in the sky seem to go round us once in 24 hours, with the Sun rising in an easterly direction and setting toward the west. It would be quite unfair of us to smile at the Ancient Chinese, Egyptians and Babylonians, who did at least make accurate observations of the skies even if they were unable to interpret what they saw. (The Egyptians believed the vault of the sky to be formed by the body of a goddess with the rather appropriate name of Nut.) The stars were divided up into constellations, and satisfactory calendars were worked out. Astronomy was all-important in timekeeping, and this was particularly vital in Egypt, where the whole economy of the country depended upon the annual flooding of the Nile. The Egyptians timed this by observations of the stars, particularly Sirius, the brightest star in the sky. When they could first see Sirius in the dawn sky, rising before the Sun, they knew when the Nile Flooding was due. Incidentally, there is no doubt at all that the famous Pyramids are astronomically aligned. We will have more to say about this later,

when we come to consider the phenomenon known as the precession of the equinoxes.

True astronomical science began with Ancient Greece. It may be said that the period of Greek astronomy began with the first great philosopher, Thales of Miletus, who was born in 624 BC; it ended with the death of Ptolemy of Alexandria, about AD 180. To tell the whole story would mean writing a separate book, so that all we can do here is to select a few of the highlights.

The first major breakthrough was the discovery that the Earth is a globe. One man who was well aware of this was Aristotle (384-322 BC), whose arguments were as powerful as they were simple. Consider the Pole Star, which lies at the northern point of the sky and seems to remain almost still, with everything else revolving round it once in 24 hours. From Greece, where Aristotle lived, the Pole Star is quite high up; from Egypt it is lower down, and there are various southern stars, such as the brilliant Canopus, which can be seen from Egypt but not from Greece. This sort of thing is easy to explain if we take the Earth to be a globe, but it cannot be explained at all on the theory that the world is flat. Aristotle used various other arguments as well, but this one alone was conclusive.

About 270 BC another great philosopher, Eratosthenes of Cyrene, managed to measure the size of the Earth. He was in charge of a vast library of books in Alexandria, and from one of these he learned that at noon on midsummer day the Sun is exactly overhead as seen from the town of Syene; at that moment the Sun's rays shine directly down to the bottom of a well without casting any shadow. Syene (the modern Aswan) lies almost due south of Alexandria, and at the same moment Eratosthenes found that from Alexandria the Sun is 7½° from the *zenith* or overhead point. All that he had to do was to measure the distance between Alexandria and Syene. A full circle contains 360°, and 7½ is about 1/50 of 360, so that if the Earth is a globe its circumference must be 50 times the distance between Alexandria and Syene; Fig. 1 should make the situation clear. Eratosthenes gave the circumference as 39,984 km, which is remarkably near the truth. We cannot be sure exactly how accurate he was, because he gave the distance in 'stadia', and there is some doubt as to the exact length of one stadion; but at any rate his value was much better than that used by Christopher Columbus in the year 1492, when he set out across the Atlantic. (Under the circumstances, it is not surprising that in searching for India, Columbus found the New World instead. The globe is much larger than he had thought.)

The next breakthrough should have been the realisation that the Earth is a planet moving round the Sun. One or two of the Greeks, notably Aristarchus of Samos (310-250 BC) did suggest this, but they could give no definite proof, and the later Greeks went back to the old idea of a central Earth. This was a pity, but in retrospect it is not surprising, even if the Earth-centred or *geocentric* theory did hold up the progress of astronomy for well over a thousand years.

Consider Hipparchus of Nicaea, who lived about 150 BC. We know virtually nothing about his life, but clearly he was a genius by any standards, and he compiled a star catalogue which was amazingly good in view of the fact that he had no telescopes. He did his best to measure the sizes and distances of the Sun and Moon, and he arrived at values

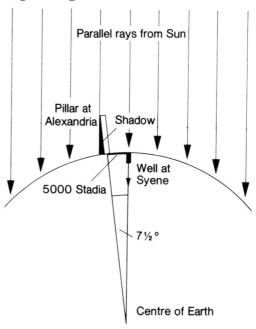

1. Eratosthenes' method of measuring the size of the Earth. The rays from the Sun can be assumed to be parallel. When the Sun shone vertically at Syene, it was 7½° from the zenith at Alexandria. The distance between the two towns was 5000 stadia. Therefore, 7½/360 = 5000/c, where c is the Earth's circumference. It worked out to 39,984 km.

which were very reasonable even though they were much too small. He also discovered the phenomenon of *precession*, a slight yearly shift in the position of the pole of the sky.

Finally, in the story of Greek science, we come to Ptolemy – Claudius Ptolemaeus of Alexandria, who lived and worked at Alexandria between about AD 120 and 180. Again we know nothing about his life or personality, but we owe a great deal to him, and periodical efforts to discredit him have been singularly unsuccessful. He wrote a great book, known to us by its Arab title of the *Almagest*, which was in effect a summary of the scientific knowledge of the time, and has been of inestimable value to historians. It contained his revision and extension of Hipparchus' star catalogue, and it also contained a description of the theory of the universe which is always called the Ptolemaic system even though Ptolemy himself did not actually invent it.

Ptolemy knew, of course, that the stars seem to remain in practically the same positions for year after year, century after century, so that to the naked-eye observer at least the constellation patterns look permanent. On the other hand the Sun and Moon shift around from one constellation to another, and so do the planets, of which Ptolemy knew five: Mercury, Venus, Mars, Jupiter and Saturn. What Ptolemy did not know, and could not reasonably be expected to know, was that the Earth itself is a planet moving round the Sun in the same way as the rest.

According to the Ptolemaic system, the Earth lies at rest in the middle of the universe. Round it move, in order of distance, the Moon, Mercury, Venus, the Sun, Mars, Jupiter, Saturn, and the sphere of the fixed stars (Fig. 2). It had always been thought that the movements of all celestial bodies must be circular, because the circle is the perfect form, and nothing short of perfection can be allowed in the heavens. However, Ptolemy, who was an excellent observer, knew quite well that the idea of uniform motion in a circular orbit did not fit the facts, because the planets behave in a much less regular way; sometimes they seem to stop and move 'backwards' against the stars for a while before resuming their ordinary behaviour. Therefore, Ptolemy preferred a complex system according to which a planet moves in a small circle or *epicycle*, the centre of which (the *deferent*) itself moves round the Earth in a perfect circle. The final theory was clumsy and artificial, but it did fit the facts as Ptolemy knew them, and for many centuries it remained more or less

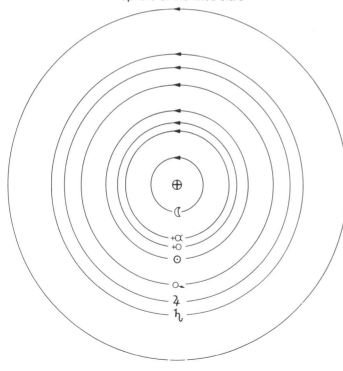

Sphere of the fixed stars

2. The Ptolemaic System. The Earth is in the centre; round it move the Moon ☾, Mercury ☿, Venus ♀, the Sun ☉, Mars ♂, Jupiter ♃ and Saturn ♄, beyond which lies the sphere of the 'fixed stars'.

unchallenged (Fig. 3).

In fact, little was added for a long time after Ptolemy's death, and astronomy – like other sciences – was in a state of hibernation. Then, in the eighth century AD, came the start of the Arab school. Ptolemy's great book was translated into Arabic (fortunately for us, as the Greek original has since been lost), and serious observing began again. The Caliph Al-Mamun founded an observatory and also a library at Baghdad around the year 830, so that Baghdad became to all intents and purposes the astronomical centre of the world. Other famous Arab astronomers were Al-Battani (around 850-929) and Al-Sûfi (around 903-986), who drew up a star catalogue which was decidedly better than Ptolemy's. Of course, good star catalogues were needed by the astrologers, who had to know the positions of the 'fixed stars' as well as the movements of the planets in order to cast their horoscopes.

The last great Arab astronomer was Ulugh Beigh, grandson of the Oriental conqueror Timur (Tamerlane). In 1433 he set up an elaborate observatory at his capital of Samarkand and carried out valuable work. Needless to say, his observatory had no optical equipment – telescopes did not come upon the scene for another 175 years – but very accurate measurements were made with equipment used with the naked eye. Unfortunately Ulugh Beigh was murdered on the orders of his son, whom he had banished on astrological advice, and with his death the Arab school of astronomy came to an end.

It was in the sixteenth century that a real breakthrough came. The man who sparked off the great controversy was a Polish churchman, Mikolaj Kopernik, usually known to us by his Latinised name of Copernicus. He was not an observer, but he was deeply interested in astronomical theory, and the more he looked at the Ptolemaic system the less he liked it. It was too complicated and

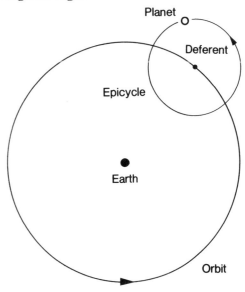

3. Epicycles. On the Ptolemaic theory, a planet moved in a small circle or epicycle; the centre of the epicycle (the deferent) itself moved round the Earth in a circle. Unfortunately even this would not account for the observed movements of the planets, and so more and more epicycles had to be introduced in an attempt to make the observations fit!

artificial. He saw that the whole scheme could be simplified merely by removing the Earth from its proud central position in the planetary system, and putting the Sun there instead. By the early 1530s he had worked out a proper *heliocentric* or Sun-centred theory, known as the Copernican system (Fig. 4), but he was reluctant to publish it, because he guessed – quite rightly – that the Church would regard it as heretical. His great book *De Revolutionibus Orbium Coelestium* (Concerning the Revolutions of the Celestial Bodies) did not appear in print until 1543, just as Copernicus himself lay dying.

Copernicus had every reason to be nervous. The Church attitude was bitterly anti-Copernican, and one scholar, Giordano Bruno, was actually burned at the stake in Rome in 1600 partly (though not entirely) because he insisted on teaching the Copernican system instead

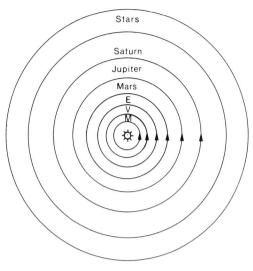

4. The Copernican System. The Sun is in the middle, but the planetary orbits are still perfect circles.

of the Ptolemaic. Also, Copernicus had by no means solved all the problems. He was still sure that all orbits must be perfect circles, and he was even reduced to bringing back epicycles. His one master-stroke was to recognise the central position of the Sun.

Ironically, the next character in the main story was an anti-Copernican. His name was Tycho Brahe; he was a Danish nobleman, and a superbly accurate observer – much the best of pre-telescopic times. Between 1576 and 1596 he worked away at his observatory on the island of Hven, in the Baltic, and he produced a magnificent star catalogue, together with precise positions of Mars and the other planets as they moved around the sky.

Tycho was nothing if not colourful. As a baby he was kidnapped by his uncle; during his student days he had part of his nose sliced off in a duel, so that he had to make himself an artificial one; he was conceited and sometimes cruel. His observatory had no telescope, but it did have a prison in which Tycho confined his

tenants who refused to pay their taxes; his retinue was enlivened by the presence of a pet dwarf named Jep. It is not surprising that Tycho was hated by the people of Hven, and eventually he quarrelled with the Danish court, so that he ended his days as a voluntary exile in Prague in Bohemia. He could not accept the heretical idea that the Earth moves round the Sun, and preferred a weird hybrid system according to which the planets moved round the Sun while the Sun itself was in orbit round the Earth. Certainly he had no patience with Copernicus. When he died, in 1601, he left all his observations to his last assistant, a German named Johannes Kepler, in the hope that they would be used to confirm that the Earth is the most important body in the universe.

Kepler lost no time in setting to work. He had complete faith in the observations of the eccentric Dane, but he soon found that the movements of Mars, in particular, could not be explained either by Ptolemy's theory or by that of Copernicus. Eventually he found the answer. The planets do indeed move round the Sun, but their orbits are not circular. Each planet moves in an ellipse; the Sun occupies one focus of the ellipse, while the other focus is empty (Fig. 5). This led him on to formulate his three famous Laws of Planetary Motion, about which we will have more to say later.

By Kepler's death, in 1630, the old Ptolemaic theory was very much on the wane, but it was far from dead, and the Church still supported it. One man who found this out, to his cost, was Galileo Galilei, the great Italian scientist who was the first astronomer to make systematic use of the telescope. (He was also the true founder of what we now call experimental mechanics.) Galileo, professor of mathematics first at the University of Pisa and then at Padua, became a Copernican early in his career, but it was not until 1610 that he became world-famous – or, in the opinion of the

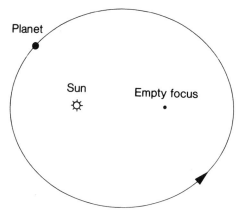

5. A planet moves in an elliptical orbit; the Sun lies at one focus of the ellipse, while the other is empty.

Church, notorious. Telescopes had just been invented, by Hans Lippershey in Holland, and Galileo made one for himself. He was not the first telescopic observer (Thomas Harriot, one-time tutor to Sir Walter Raleigh, used a telescope to map the Moon some months before Galileo heard about the Dutch invention), but for skill and patience Galileo far outshone any of his contemporaries.

Over a period of a few months, Galileo made a series of spectacular discoveries. He saw the mountains and craters of the Moon, and found that the band of light across the night sky which we call the Milky Way is made up of vast numbers of faint stars. He discovered that the planet Jupiter is attended by four satellites, so that clearly there had to be more than one centre of motion in the Solar System. Even more significantly, he found that the planet Venus shows phases, or apparent changes of shape, similar to those of the Moon – which on the Ptolemaic theory it could never do.

Less tactful than Copernicus or even Kepler, Galileo made no secret of his views, with the inevitable result that he was called to Rome, brought to trial by the Inquisition, and forced to 'curse, abjure and detest' the false theory that the Earth moves round the Sun. He was not tortured, but he ended his days under house arrest in his villa; he died in 1642. It is a sad story, and it is very much to the discredit of the Church. (He was finally given an official pardon in 1992!)

Yet Galileo was the last great scientist to be persecuted for 'heresy' of this kind, and before the end of the seventeenth century the Ptolemaic theory was a thing of the past. It was finally killed off by the work of Sir Isaac Newton, who laid the foundations of modern-type astronomy.

Newton has been described as the most brilliant mathematician who has ever lived. Whether this is true or not, he can have had few equals. He studied at Cambridge, and in 1665-6, when the University was temporarily closed because of the danger of plague, he stayed at his home at Woolsthorpe, in Lincolnshire, carrying out researches of fundamental importance. Years later, in 1687, he published his *Philosophiae Naturalis Principia Mathematica*, which is always known by its shortened title of the *Principia*. It was here that he first published the laws of universal gravitation, according to which every body attracts every other body with a force which becomes weaker with increasing distance.

Newton's work extended into many fields. He built the first reflecting telescope, which collects its light by means of a mirror rather than a glass lens; he investigated the nature of light, and proved that what we call 'white' light is really a mixture of all the colours of the rainbow (a discovery which led on to the development of the spectroscope), and he was an independent inventor of the calculus, which has become so vital in all mathematics. In his later life he became Master of the Royal Mint, and played an important part in revising Britain's coinage, but it is as a scientist that he will always be remembered.

The genius of Newton tended to overshadow the other brilliant astro-

11

6. Edmond Halley, the second Astronomer Royal.

Flamsteed had the advantage of being able to use telescopes, and the catalogue was eventually produced, though it took many years to complete; the final version was not published until after Flamsteed had died.

Flamsteed was the first Astronomer Royal. The second was Edmond Halley, who was a close friend of Newton's, and actually paid for the publication of the *Principia* (indeed, without Halley's constant urging, the book would probably never have been written). Halley is best remembered nowadays for his successful prediction of the return of a comet (about which we will say more later), but he made many other contributions as well – such as the discovery that several of the brilliant stars had shifted slightly in position since the time of Ptolemy. And when discussing astronomers of the late seventeenth century we must not omit Sir Christopher Wren, who was professor of astronomy at Cambridge before turning to architecture, and was a first-rate mathematician. It was Wren who designed the original Royal Observatory at Greenwich; the buildings are still there, though they have now been turned into a museum.

There were other great astronomers living in Europe at the same time. For instance Ole Rømer of Denmark measured the velocity of light, and Giovanni Cassini, an Italian who spent much of his life in France and was the first Director of the Paris Observatory, made a series of telescopic discoveries, including the main gap in Saturn's rings and several of the planet's satellites. But in a brief historical outline we must now skip almost a century, and come next to William Herschel, whose astronomical career stretched from 1774 up to his death in 1822.

Herschel was born in Hanover, but came to England while still a young man, and became a music teacher and performer. For a time he was organist at the Octagon Chapel in Bath, and earned

nomers of the time, notably Robert Hooke, John Flamsteed, Edmond Halley and Christopher Wren. Hooke, for example, was remarkably versatile, and as a mathematician was probably second only to Newton himself – though by all accounts he had a difficult personality, and it is certainly true that Newton refused to publish the second of his great books, dealing with optics, until after Hooke's death.

Flamsteed was above all an observer, and he was put in charge of the new observatory that was set up at Greenwich, in 1675, by the express order of King Charles II, so that the stars could be re-catalogued for use by British sea navigators. This had become an urgent need, because sailors far out of sight of land found it difficult to measure their longitude. The only method that seemed to be suitable involved measuring the position of the Moon against the stars, but an accurate catalogue was essential, and even Tycho's was not good enough.

7. The Old Royal Observatory at Greenwich, as photographed in 1988. Almost central is the open door of the Airy Transit Circle, which marks the zero for longitude.

himself a wide reputation. When he became interested in astronomy he decided to make his own telescopes, and he built reflectors which were much the best of their time. With one of these, in 1781, he discovered a new planet, now known as Uranus. It moves round the Sun far beyond the orbit of Saturn, which was the most remote planet known in ancient times, and it is barely visible with the naked eye.

This discovery altered Herschel's whole career. A modest grant from King George III of England and Hanover made it possible for him to give up music as a profession, so that he could devote his whole time to astronomy – in which he was helped by his devoted sister Caroline, herself an excellent observer who discovered several comets.

Using his powerful telescopes, the largest of which had a main mirror 124 cm across, Herschel undertook systematic 'reviews of the heavens', discovering and cataloguing thousands of double stars, star clusters, and the dim, misty patches which we call nebulae,

some of which are clouds of gas and dust while others are now known to be galaxies in their own right. There was no branch of observation which Herschel did not touch, and he has been rightly called 'the father of stellar astronomy'. As an observer he has never been equalled, even though some of his ideas sound strange today; he believed that the habitability of the Moon was 'an absolute certainty', and he even thought that there were intelligent beings living in a cool region below the brilliant surface of the Sun!

Herschel set out to measure the shape and form of the Galaxy. This he did by counting the numbers of stars visible in carefully-selected areas of the sky. Eventually he decided that the system was shaped like a 'cloven grindstone', though it is often likened to the shape of two fried eggs clapped together back to back. Basically Herschel was right, though he was wrong in placing the Sun near the centre of the system; actually it

8. Wedgwood representation of William Herschel (reproduced by kind permission of Wedgwood Ltd).

13

is almost 30,000 light-years away from the galactic centre.

One task, however, defeated him. It had long been known that the stars are suns, and that the other stars are much more remote than our Sun (remember that the Earth-Sun distance is 1 astronomical unit, which is approximately 150,000,000 km). Herschel did his best to measure the distance of the stars. He selected pairs of stars (doubles) because he believed that one member of the pair would be much closer than the other, so that the second star would be in the background, so to speak; in this case there should be a regular apparent shift due to the Earth's motion round the Sun. In fact, no such shift was detected, because most double stars are physically-associated or *binary* pairs, moving through space together and at the same distance from us. It was not until 1838, sixteen years after Herschel's death, that the German astronomer F.W. Bessel managed to make the first measurement of the distance of a star. By the method of *parallax*, about which we will have more to say later, he established that 61 Cygni, a dim star in the constellation of the Swan, is approximately 11 light-years away, corresponding to around 106 million million km.

Just as Newton overshadowed his contemporaries of the late seventeenth century, so the achievements of Herschel tend to mask the talents of his fellow astronomers of the late eighteenth century. Yet there were observers and theorists in plenty. Johann Schröter, of Lilienthal, laid the foundations of the physical study of the Moon. Charles Messier, in France, drew up the first proper catalogue of star-clusters and nebulae, not because he was interested in them, but because he kept on confusing them with comets – in which he was very interested indeed (altogether he discovered more than a dozen). Then there was a most unusual astronomer, John Goodricke of York, who was born deaf and dumb and who died when he was only twenty-one, but who made some remarkable discoveries in connection with variable stars; he suggested, quite correctly, that some stars which change in light, such as the 'Demon Star' Algol in Perseus, are not truly variable at all, but are binary systems in which one member of the pair is brighter than the other – so that when the fainter star passes in front of the brighter, the total brilliancy, as seen from the Earth, drops. On the theoretical side, Pierre Simon de Laplace, of France, discussed the origin of the Earth and the other planets, putting forward a scheme which is not so very unlike some of our modern ideas even though it has needed drastic modification.

From the beginning of the nineteenth century, the history of astronomy becomes much more complicated, because so many completely new branches were being developed. First came *spectroscopy*. To be fair, it was really sparked off by Newton's experiments at Woolsthorpe in the mid-1660s, when he used a prism to split up the rays of the Sun and showed that they could be spread out into a rainbow, but it was only after 1800 that anything more was done. Note that it was William Herschel who first realised that when he spread out the sunlight into a band or spectrum (Fig. 9) he could still detect 'heat' beyond the red end of the band – this was the first direct observation of what we now call infra-red.

To anticipate what we will be saying in more detail later: light is a wave-motion, and the colour of light depends upon its wavelength, red light having the longest wavelength and violet the shortest, with orange, yellow, green, blue and indigo in between. By splitting up light in this way we can tell what substances are present in the light-source. In 1814 the German optician Josef von Fraunhofer studied the spectrum of the Sun, and found that the rainbow band was crossed by dark lines (Fig. 67). We now know that each of these dark lines is due to some particular

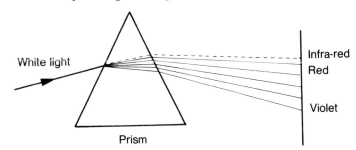

9. Herschel's discovery of infra-red radiation.

element, such as hydrogen, oxygen or iron, so that we can find out 'what the Sun is made of'. It is interesting to recall that even in 1830 a French philosopher named August Comte was writing that mankind could never hope to find out anything about the chemistry of the stars. (This may perhaps demonstrate that when a French philosopher makes a profound statement, he is almost certain to be wrong!)

In the 1830s spectroscopy was joined by photography. Indeed, the very word 'photography' was coined by an astronomer – Sir John Herschel, son of Sir William, who also achieved fame by travelling to the Cape of Good Hope where he stayed between 1833 and 1838, making the first systematic survey of the far-southern stars which never rise over Europe. Of course, early photographic processes were very crude by modern standards, but by the end of the century photography had replaced the human eye for most branches of astronomy, so that the world's largest telescopes were used mainly as giant cameras.

In 1845 the third Earl of Rosse, an Irish nobleman, built what was then much the largest telescope ever made. It was a reflector with a 183-cm mirror, and it was set up at Birr Castle, in Central Ireland. It was an extraordinary instrument, with a very limited view of the sky (it was mounted between two massive stone walls) but it worked well, and with it Lord Rosse discovered that some of the dim misty objects catalogued by Messier and Herschel were spiral in form, like Catherine-wheels. At the time, all that could really be said was that the spirals were different in nature from the formless nebulae which looked like (and were) clouds of dust and gas. It was almost eighty years before it could be shown that the spirals are independent galaxies, millions of light-years away. The Rosse telescope has now been restored, and is again operational.

There was one major theoretical triumph in 1846. Uranus, the planet discovered by Herschel, had been wandering away from its predicted path; clearly some force was pulling it out of position, and two mathematicians, John Couch Adams in England and Urbain Le Verrier in France, independently decided that the disturbing force must be due to an unknown planet moving still further out from the Sun. They calculated where this planet should be, and, sure enough, the world we now call Neptune was found in almost exactly the position which they had predicted.

15

Towards the end of the century, various large telescopes were built both in Europe and in America. Most of them were refractors, collecting their light by means of large lenses known as object-glasses or objectives. Among them were the 91.4-cm refractor at the Lick Observatory in California, the 83-cm refractor at Meudon in France, and the 102-cm or 40-inch refractor at the Yerkes Observatory, not far from Chicago. The Yerkes telescope was the brainchild of George Ellery Hale, an American who began his career as an amateur concerned mainly with studies of the Sun, and who had the useful knack of persuading friendly millionaires to finance his projects (something which was much easier a hundred years ago than it is today). In this case the millionaire was C.T. Yerkes, a local business magnate. But Hale was not satisfied; he realised, too, that there is a practical limit to the size of a refractor. If an object-glass is too large, it will also become so heavy that it will distort under its own weight, making it useless, so that the Yerkes refractor remains the largest of its kind. (It is still in use on every clear night; I observed Mars with it not long ago – P.M.)

Instead of persevering with refractors, Hale turned his attention to reflectors. A mirror is supported at its back, and so there is no limit to size except that imposed by the unsteadiness of the atmosphere and the sheer mechanical problems of mounting. With the support of another millionaire, J.D. Hooker, Hale set up a 152-cm (60-inch) reflector on Mount Wilson in California; the mountain site was chosen because the higher the altitude, the thinner the atmosphere overhead. Next came the 254-cm (100-inch) Hooker reflector, also at Mount Wilson, which was for decades the most powerful telescope in the world, and was indeed in a class of its own. It was with this telescope, in 1923, that Edwin Hubble made the observations that proved that the spirals and other 'starry nebulae' really are external galaxies rather than mere parts of our own Milky Way.

Hale also planned an even larger reflector: the 508-cm (200-inch) telescope at Palomar Mountain in California. Unfortunately he did not live to see it completed, but when it came into use, in 1948, it revolutionised astronomy. With its tremendous light-grasp it could 'see' further into space than ever before, and, like the Mount Wilson telescope, it remained in a class of its own for many years.

One ever-increasing problem was atmospheric pollution – both industrial haze and the brilliance of artificial lights. The Mount Wilson reflector was even taken out of commission during the 1980s because of the glare from the city of Los Angeles, and although it has now been brought

10. The Hale 200-inch reflector at Palomar Observatory, California, as photographed in 1986.

back into full use there is no doubt that conditions are much less favourable than they were in Hale's day.

These were also problems facing the Royal Greenwich Observatory, which had a unique reputation and had also been accepted as the 'timekeeping centre' of the world (the zero for longitude, dividing the globe into its eastern and western hemispheres, passes through one of the Observatory's instruments which had been set up by Sir George Airy, Astronomer Royal for many years during the nineteenth century). With the spread of London, Greenwich Park became unsuited to observation of any kind, and in the 1950s the Observatory was moved to Herstmonceux Castle in Sussex where a 244-cm (98-inch) reflector, the Isaac Newton Telescope or INT, was installed. But Herstmonceux itself was far from ideal, and in 1983 the INT was moved to a new site, La Palma in the Canary Islands. This was one factor leading to the eventual closure of Herstmonceux, and the Royal Greenwich Observatory was itself closed down – to the great regret of astronomers everywhere.

La Palma is an excellent site, and the INT has been joined by an even larger reflector, the William Herschel Telescope or WHT, which completed its tests in 1988 and has proved to be superb in every way. There are many other telescopes at La Palma, too, made and operated by various nations, notably Denmark, Holland and Sweden. The La Palma observatory, known officially as Los Muchachos, has become truly international.

All the first major observatories were in the northern hemisphere. However, many of the most interesting objects in the sky lie in the far south, so that from Europe and the United States they never rise. For this reason there has been a policy of setting up most of the new large observatories in the southern hemisphere, or at least far enough south to cover most of the sky. In 1974 came the 389-cm Anglo-Australian Telescope

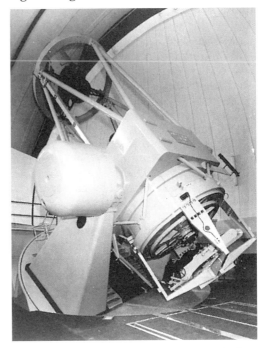

11. The Isaac Newton telescope at La Palma, in the Canary Islands.

(AAT) at Siding Spring in New South Wales. The summit of the extinct volcano Mauna Kea, in Hawaii, has become an astronomical centre; at an altitude of 14,000 feet (over 4200 metres) the atmosphere is thin and clear. Another excellent site is Chile, where there are now four leading observatories: La Silla with a 381-cm reflector, Cerro Tololo with a 401-cm, Las Campanas with a 254-cm, and the new VLT. Though on Chilean soil, La Silla and VLT are run by

12. Dome of the William Herschel Telescope, at La Palma, Canary Islands.

the Europeans and the other two by the Americans. In the United States itself we have Kitt Peak in Arizona, which has a 401-cm reflector as well as the world's largest instrument designed specially for studying the Sun.

The largest single-mirror telescope has a 600-cm mirror. It was made in the Soviet Union, and was brought into use in 1976. It cannot really be said to have come up to expectations, possibly because the site in the Caucasus Mountains is not nearly so good as the peaks of Hawaii, La Palma or Chile, but it is historically important in one particular way: it has a new type of mounting, and is completely controlled by computers. This design has worked well, and all later large telescopes have mountings of the same kind.

So far we have been discussing single-mirror telescopes, but the modern trend is to use mirrors made in segments which can be fitted together to produce the correct optical curve, or to use multiple-mirror telescopes. In 1991 the Keck telescope was completed, with a segmented mirror 9.8 metres in aperture; it was set up on Mauna Kea, and was subsequently joined by Keck II, which is identical – the two can be used together, giving them immense power. Also on Mauna Kea are the Japanese Suburu telescope with an 8.3-metre mirror, and the Gemini North telescope, with an aperture of 8.3 metres; a second Gemini (South) is being constructed, and will be set up in Chile. At present (1999) the most powerful telescope in the world is the VLT or Very Large Telescope, at Cerro Parañal in the northern Atacama Desert. When completed, it will have four 8-metre mirrors working together; two of these were operating before the end of 1999.

Another major development is what we usually call the 'Electronic Revolution'. Just as the photographic plate replaced the human eye a century ago, so photography is itself now being superseded by electronic devices, which are far more sensitive. Of special importance is the CCD, or Charge-Couple Device. This is a silicon integrated circuit consisting of an oxide-covered silicon substrate upon which is formed an array of closely-spaced electrodes. This is no place to go into detail; it is enough to say that telescopes working with CCD equipment are much more effective than they could ever be with photographic plates. The 'old' telescopes, such as the Palomar reflector, have much greater range now than they used to have.

A few decades ago an astronomer studying, say, a remote galaxy had to spend many hours in the observatory, sitting at the eye-end of a telescope and guiding it. Today the situation is quite different. Everything is computerised; the observer does not stay in the dome – he is in a separate control room, watching a television screen. He need not be in the observatory at all, or even in the same country. It is quite practicable to stay in a control room in Britain and operate a telescope on Mauna Kea or La Palma, and

13. The Lovell radio telescope at Jodrell Bank, as photographed in 1988.

18

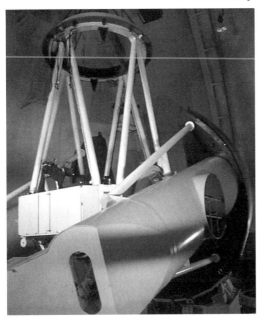

14. UKIRT, the United Kingdom Infra-Red Telescope on Mauna Kea, Hawaii.

as time goes by this policy of remote control is becoming more and more common.

Moreover, astronomy is an international science, and the International Astronomical Union or IAU, the controlling body of world astronomy, really lives up to its name.

15. Domes atop La Silla in the Atacama Desert, Northern Chile.

There are no Iron Curtains, or even Bamboo Curtains, among astronomers.

We have run somewhat ahead of our story, and we must go back now to say something about what is often called 'invisible astronomy'. As we have seen, William Herschel was the first to prove that the Sun sends out infra-red radiation as well as visible light, and today infra-red astronomy has become highly important. Because much of the infra-red radiation from the sky is blocked out by the Earth's air, particularly by the water-vapour content, it is essential to observe from high altitudes; the world's largest telescope designed for infra-red work, the United Kingdom Infra-Red Telescope, UKIRT, has been set up on the summit of Mauna Kea, Hawaii, above most of the atmospheric water-vapour.

In 1931 Karl Jansky, an American radio engineer, was using an improvised aerial to study 'static' when he found that he was picking up long-wavelength radiations from the Milky Way. This was the start of radio astronomy, which has today become one of the main branches of research; who has not heard of the 250-foot (76-metre) 'dish' at Jodrell Bank in Cheshire, now named the Lovell Telescope in honour of its creator, Sir Bernard Lovell? Radio telescopes are to be found in every country; in Puerto Rico there is even a vast 'dish' built in a natural hollow in the ground.

There are still some people who fondly imagine that radio astronomy has 'taken over' from optical astronomy. Of course this is not so; the two lines of research are complementary. To all intents and purposes a radio telescope is a huge aerial, and one certainly cannot look through it, but it can provide us with information which we could never obtain in any other way.

Since 1957, when the Russians launched the first artificial satellite, Sputnik 1, astronomers have been able to investigate the universe in a whole multitude of ways. Of course, the best known programme is that of the Apollo landings on the Moon, which, although mainly made for political reasons, brought back much good science and told us more about the lunar world as a whole. Ever since, the emphasis has been more on robotic exploration of the Solar System, and all the major planets with the exception of Pluto have now been visited. Voyagers 1 and 2 revealed the giant outer planets in all their glory, various probes explored the surface of Mars; Giotto managed to rendezvous with Halley's Comet. Visits were also made to Venus, with various pioneering Russian probes landing on the surface before the American Magellan provided us with full radar mapping of the surface. Closer to home, Shuttle missions have enabled the launch (and later repair) of observatories such as the Hubble Space Telescope, which has truly revolutionised our view of the universe. Many of the pictures and discoveries from these missions are reported in more detail later in the book.

We realise that this has been a very sketchy and incomplete review, but we do not want to fall into the trap of the 'penguin syndrome', and we hope we have at least given you the main essentials. Henry Ford was wrong in claiming that 'history is bunk'. Unless we have some idea of the way in which our knowledge has been built up, we will be unable to form a proper appreciation of that knowledge itself.

16. This 210-foot radio antenna at the Goldstone, California complex of tracking stations, is the prime instrument in a world-wide network of deep space stations. Primarily used for communications with unmanned planetary spacecraft, and Apollo landers, the instrument is also used for radio and radar astronomy research.

Questions

1. Give one reason why Aristotle believed that the Earth is a globe rather than a flat plane.
2. What was the main contribution made by Copernicus to astronomical knowledge?
3. What did Johannes Kepler discover about the shapes of the orbits of the planets?
4. Explain the main reason why Greenwich Observatory was founded in 1675.
5. Herschel set out to determine the shape and form of the Galaxy. What was his main conclusion about this?
6. Explain how it was possible to work out the position of Neptune, in 1846, before the planet was actually seen.
7. Why can a reflecting telescope be made to a larger aperture than a refractor?
8. What important discovery was made by Karl Jansky in 1931?
9. Give two advantages and two disadvantages of positioning a telescope above the Earth's atmosphere.
10. Discuss some of the advantages of international co-operation in astronomy.

4
The Earth as a Planet

Because the Earth is our home, we tend to think of it as being particularly important. Of course it is important to us, but it is a very ordinary planet, and there are not many ways in which it can be singled out from the other members of the Sun's family.

In size it is normal enough. It has a diameter of 12,756 km as measured through the equator, but only 12,709 km as measured through the poles, so that it is not a perfect sphere; it is what is known technically as an oblate spheroid (Fig. 17). This is because it rotates on its axis, and this makes the equator bulge slightly. As Kepler showed, the Earth's orbit is not circular; it is an ellipse, with the Sun lying in one focus. At its closest to the Sun (*perihelion*) the separating distance is 147,000,000 km; at its furthest (*aphelion*) it is 152,000,000 km, giving a mean of approximately 150,000,000 km – to be more precise, 149,598,500 km, or 1 *astronomical unit*. The time taken to complete one orbit is 365.256 days. This is slightly longer than our calendar year of 365 days, which is why we have to add an extra day every four years to make our calendar correct; in Leap Years, February has 29 days instead of its usual 28.

Yet there are some ways in which the Earth is different from the other planets. It is the only world in the Solar System which is suitable for advanced life of our sort. It is the right distance from the Sun, it has the right kind of temperature and atmosphere, and it alone has oceans of water; approximately three-quarters of the surface is covered with water.

The atmosphere is unlike any other

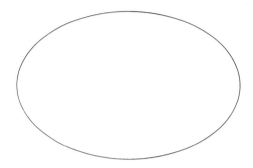

17. An oblate spheroid. For the sake of clarity, the flattening is very much exaggerated compared with that of the Earth!

known to us. In composition it is made up of 78 per cent nitrogen and 21 per cent oxygen, which does not leave much room for anything else, though there are appreciable quantities of other gases such as argon and carbon dioxide. It extends upward for hundreds of kilometres, but most of it is concentrated in the lowest layer, known as the troposphere, which lies between sea-level and a mean height of 8 km over the poles and 11 km over the tropics. This is where we find all our 'weather', and all ordinary clouds. The atmosphere thins out with increasing height; there is no definite boundary, and the air simply 'tails off' until its density is no greater than that of the interplanetary medium. It is fair to say that there is very little trace of atmosphere left above a height of 1000 km.

The atmosphere is unsteady, and this is what makes the stars twinkle – their light is, so to speak, shaken about as it comes through the air. A star is to all intents

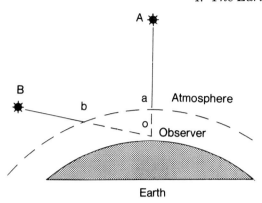

18. Scintillation. The observer is at o; the dashed line represents the Earth's atmosphere. Star A, high above the observer's horizon, twinkles less than star B, because the distance ao, through which the light has to pass, is less than the distance bo.

and purposes a point source, and when low down its light comes to us through a thick layer of atmosphere, so that it twinkles (*scintillates*) violently; when it is high up, it twinkles less (Fig. 18). Water vapour in the high part of the troposphere may split up the sunlight and cause rainbows; sometimes a thin layer of ice-crystal cloud will make the Sun and the Moon produce a halo. All these phenomena belong more properly to meteorology than to astronomy, and it is of course the atmosphere which scatters the blue part of the Sun's light around to make the sky blue. When the Sun is low down, more of its red rays can pass unchecked; most people have at one time or another admired the beauty of a red sunset.

There is no problem in understanding the cause of day and night. Quite clearly, the Sun can light up only half the Earth at once, just as a torch can light up only half a football in a darkened room, and the Earth spins round once in approximately 24 hours. Note, incidentally, that in astronomy the term 'day' is taken to mean the whole 24-hour period, not just the interval between sunrise and sunset.

All of us are familiar with the march of the *seasons* – spring, summer, autumn and winter – but these have very little to do with the Earth's changing distance from the Sun, and we are actually at our closest to the Sun in December, when it is winter in Britain. The cause of the seasons is quite different.

The Earth's axis is tilted to the plane of its orbit round the Sun; the angle between the axis and the perpendicular is 23°26', or approximately 23½°. This is the same as saying that the angle between the orbital plane and the Earth's equator is 23½°. In June, the Earth's north pole is tilted toward the Sun, and the northern hemisphere receives the full benefit of the Sun's rays; the north pole is in daylight all the time, while the south pole has no sunlight at all. By December the conditions have been reversed. Fig. 19 should make the situation clear.

The Sun moves round the sky once a year, but it does not move along the celestial equator. It reaches its northernmost point around 22 June (*summer solstice*) and its southernmost point around 22 December (*winter solstice*). It crosses the equator of the sky twice, once around 21 March, when passing from south to north (*spring* or *vernal equinox*) and again around 22 September, when moving from north to south (*autumnal equinox*). Only at the equinoxes are days and nights equal in length in the two hemispheres of the Earth.

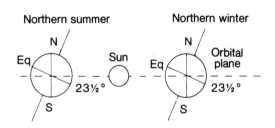

19. The seasons. In northern summer, the North Pole is tilted toward the Sun. The angle between the equator and the orbital plane is 23½°; in other words, the Earth's axis is inclined by 23½° to the perpendicular to the orbital plane.

The changing distance between the Earth and the Sun must have some effect on the seasons. Winter in the southern hemisphere should be longer and colder than in the north, while the summer should be shorter and hotter. In theory this is true enough, but the difference in distance is not great – only about 5,000,000 km – and there is much more ocean in the Earth's southern hemisphere, which tends to stabilise the temperature there.

When we want to give the position of any point on the Earth's surface, we do so by quoting its latitude and longitude. *Latitude* is simply the angular distance of the point from the Earth's equator reckoned from the centre of the globe (Fig. 20); thus Selsey in Sussex is at latitude 50°44'N. Obviously, the latitude of the equator is 0°, with the north pole at 90°N (or +90°) and the south pole at 90°S (or –90°).

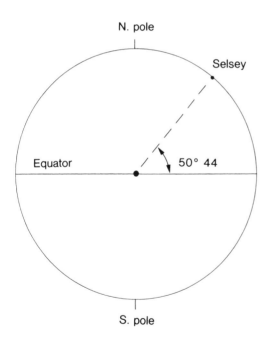

20. The latitude of a point is the angular distance from the equator, reckoned from the centre of the Earth. For Selsey in Sussex it is 50°44'N; for the North Pole it is 90°N, and for the equator it is 0°.

Longitude is the angular distance of any point east or west of the Greenwich meridian. This is straightforward enough, but it is essential to understand what a *meridian* is, and this brings in the term *great circle*, because the official definition of a meridian is 'a great circle on the Earth's surface which passes through both poles'.

A great circle is a circle drawn on the surface of a sphere, and whose plane passes through the centre of the globe, so that if you sliced the Earth through along a great circle you would cut the world in half. The experiment can be carried out easily with the help of an orange. Cut the orange in half, and then clap the two halves together again; the line marking the join will be a great circle on the surface of the orange.

The Greenwich meridian, then, is the great circle on the Earth's surface which passes through both poles and also through Greenwich Observatory – the old Royal Observatory in Greenwich Park (Fig. 21). It was accepted as marking longitude 0° over a century ago, by international agreement. Longitude is measured up to 180° east and west. The longitude line at 180° is the 'other half' of the Greenwich line of 0°, and is of course on the opposite side of the world; it runs largely through the Pacific Ocean, missing New Zealand by a fairly wide margin.

Now we can come back to Selsey, whose latitude has already been found to be 50°44'N. Its longitude turns out to be 0°48'W, so that it is practically on the Greenwich meridian. With this information, we can pinpoint exactly where our observation point lies on the surface of the Earth. (To be precise, Selsey village covers a considerable area, so that we have had to choose one particular point – and it seemed appropriate to select the observatory dome which covers P.M.'s largest telescope, a 39-cm reflector!)

24

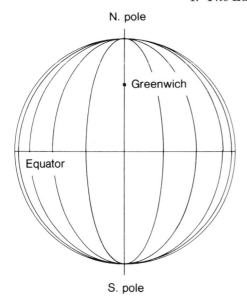

21. Longitude. The Greenwich meridian is taken as 0°. Longitude is measured to 180°E and W; obviously, 180°E is the same as 180°W.

Note that in every case we are reckoning according to the geographical poles, which are not the same as the magnetic poles. We have to admit that even now we are not confident that we know all we would like to know about the Earth's magnetic field, but we have found out how it behaves; the magnetic poles are not coincident with the geographical ones, so that a magnetic compass does not indicate true north. The difference between true north and magnetic north is termed *variation*, and is different for different places on the Earth; it also changes somewhat with time. Note, incidentally, that not all bodies have overall magnetic fields; Venus has no detectable fields, and neither has the Moon. Mars' is very weak. Mercury has an appreciable field, and the giant planets have very strong fields indeed.

Because the Earth spins on its axis from west to east, the bodies in the sky seem to move from east to west, taking approximately 24 hours to complete a full circuit. Yet there are two points in the sky which do not seem to move at all; these are the celestial poles, which lie in the direction of the Earth's axis (Fig. 22). The north celestial pole is marked fairly closely by the bright star which we call Polaris, or the Pole Star, in the constellation of Ursa Minor, the Little Bear. There is no bright star close to the south celestial pole, and we have to make do with the obscure star Sigma Octantis, which is visible with the naked eye but is not particularly easy to recognise.

For the moment, let us suppose that

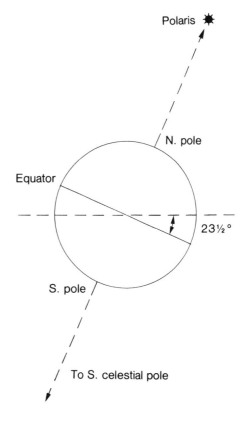

22. The celestial poles. Polaris, the Pole Star, lies almost in the direction of the Earth's axis; it is less than 1° from the celestial pole. Therefore, Polaris seems to remain almost stationary in the sky, with everything else moving round it in 24 hours. The south celestial pole, unmarked by any bright star, is of course invisible from Europe or any part of the Earth's northern hemisphere. The dashed line at 23½° to the equator is the plane of the Earth's orbit.

Polaris lies exactly at the north celestial pole instead of being slightly less than one degree away from it (its precise declination is +89°15'51"). To an observer standing at the north pole of the Earth, Polaris will be seen at the zenith or overhead point. To an observer on the equator, Polaris will lie on the horizon, and to anyone looking at the sky from a point south of the Earth's equator Polaris will never be seen at all. As Aristotle realised so long ago, one's view of the sky depends upon one's own position on the surface of the Earth. In particular, the altitude of Polaris – that is to say, its apparent height above the horizon – will alter. Altitude is measured in degrees, from 0° at the horizon up to 90° at the zenith.

Once again let us return to Selsey, with its latitude of 50°44'N. Measure the altitude of the celestial pole, marked by Polaris, and you will find it also to be 50°44'. In fact, *the altitude of Polaris is always the same as your own latitude*, a fact which air and sea navigators find very useful. Once the altitude of Polaris has been measured with a sextant, the observer's latitude is known – or, rather, it will be known as soon as allowance has been made for the fact that Polaris is not exactly at the pole. From the equator (latitude 0°) Polaris has an altitude of 0°, so that it stays on the horizon. From countries such as Australia or South Africa, the same kinds of calculations can be made by measuring the altitude of the south celestial pole, though the lack of any bright star there makes matters rather more difficult.

Though Polaris is so useful to navigators, and occupies a special position in our sky, there is nothing important about it in itself; it is simply a normal star (though it is much more luminous than our Sun). It has not always been the pole star, and will not remain so indefinitely, because of an effect known as *precession*.

As we have noted, the Earth is not a

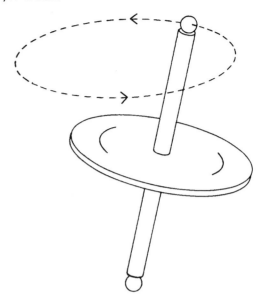

23. Precession: the same basic effect as the toppling of a gyroscope which is running down.

perfect sphere; the equator bulges out slightly. The Sun and Moon pull upon this bulge, and the effect is to make the Earth's axis wobble slightly, rather in the manner of a gyroscope which is running down and has started to topple (see Fig. 23). But whereas the gyroscope swings round in a few seconds, the Earth's axis takes 25,800 years to describe a circle in the sky.

Several thousands of years ago, when the Egyptians were building their Pyramids, the polar point lay not near Polaris, but close to a much fainter star, Thuban in the constellation of the Dragon. To the Egyptians, then, Thuban was the pole star. In 12,000 years hence it will be the turn of the brilliant bluish star Vega; in 25,800 years from now the polar point will have completed a full circle in the sky, 47° in diameter, and will be back to Polaris. Obviously, precession is a slow process, and it is all the more remarkable that it has been known ever since the time of Hipparchus in the second century BC (Fig. 24).

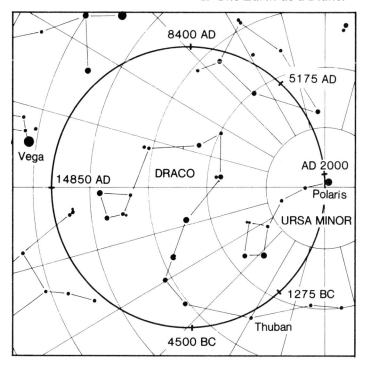

24. Precession: the shift of the celestial pole. At present the north pole of the sky lies near Polaris. In 2500 BC it lay near Thuban in Draco (the Dragon). In 12,000 years' time the pole star will be Vega. In 25,800 years' time the pole will have completed one circuit and returned to its present position near Polaris.

It is often said that the Sun rises in the east and sets in the west. However, this is not the full story. When the Sun lies exactly on the equator of the sky, at the March and September equinoxes, it does admittedly rise due east and set due west; but when the Sun is in the northern hemisphere of the sky (late March to late September) it rises north of east and sets north of west, while when it is in the southern hemisphere (late September to late March) it rises south of east and sets south of west. It follows that in the northern hemisphere, anyone who travels northward in summer will find that he is experiencing longer and longer periods of daylight, until in the polar region it will be daylight for the full 24 hours. This is why the June daylight periods in, say, Oslo or Stockholm are longer than those in London – but to make up for this, London has much more daylight in winter.

Stars which are close to the celestial pole do not set, but merely go round and round in circles, staying above the horizon all the time. These are known as *circumpolar* stars. In the left-hand diagram of Fig. 25, star A, not far from Polaris, is circumpolar from the latitude of London, approximately 51½°N; star B, further from the pole, is not circumpolar, because it spends part of its daily journey below the horizon. Go down to latitude 10°, in the right-hand diagram of Fig. 25, and even star A will not be circumpolar.

Because the Earth goes round the Sun once a year, the Sun seems to travel right round the sky once a year, moving through the constellations which make up the *Zodiac*. For instance, in late March the Sun lies among the faint stars of the constellation Pisces, the Fishes. We cannot see the Sun and the stars at the same time, for obvious reasons, so that during March Pisces is out of view. Different constellations are observable at different seasons, though groups which

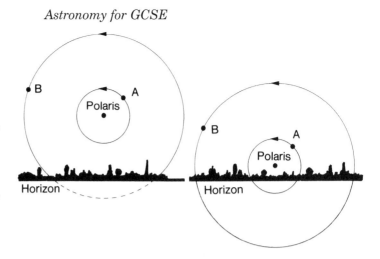

25. Circumpolar and non-circumpolar stars. Assume that we are observing from the Earth's northern hemisphere. *Left*: star A, not far from Polaris, is circumpolar; star B is not, as part of its daily circuit is spent below the observer's horizon. *Right*: from a position further south on the Earth's surface, Polaris appears lower down; now neither A nor B is circumpolar.

lie closer to the poles, well away from the Zodiac, can be seen throughout the year.* Since the Sun completes one journey round the sky (360°) in approximately 365 days, its apparent motion against the stars, travelling from west to east, is roughly one degree per day. But we must always remember that the east-to-west *diurnal* (daily) movement of the sky is due entirely to the Earth's rotation, and has nothing directly to do with either the Sun or the stars.

There is one experiment, easily made, which will demonstrate the diurnal motion. Take a camera, load it with a suitable film, and go out at night; aim the camera at the Pole Star, and give a time-exposure of, say, an hour or so. You will find that the stars appear as trails – and even Polaris shows a short curved trail, showing that it is almost a degree away from the true pole of the sky.

* The twelve Zodiacal constellations are Aries (the Ram), Taurus (the Bull), Gemini (the Twins), Cancer (the Crab), Leo (the Lion), Virgo (the Virgin), Libra (the Balance), Scorpius (the Scorpion), Sagittarius (the Archer), Capricornus (the Sea-Goat), Aquarius (the Water-bearer) and Pisces (the Fishes). A thirteenth constellation, Ophiuchus (the Serpent-bearer) intrudes into the Zodiac between Scorpius and Sagittarius, though it is not usually counted as a Zodiacal group.

Questions

1. (a) Give two reasons why the Earth is different from all the other planets in the Solar System.
(b) What is the true shape of the Earth?
(c) Explain what is meant by the terms *perihelion* and *aphelion*.
2. (a) Why does a star twinkle more when it is low down than when it is high up?
(b) Why does the Sun often look very red when setting?
(c) Why is the sky blue?
3. (a) Explain what is meant by *latitude* and *longitude*. What is (i) the latitude of the equator, (ii) the latitude of the south pole, (iii) the longitude of Greenwich Observatory?
(b) Using diagrams, give an explanation of the phenomenon of the seasons.
(c) What is meant by the term *equinox*?

4. (a) During June, two boys took their holidays at the same time, one in Norway and the other in Greece. When they compared notes, they found that the boy who had been to Norway had had more hours of daylight than his friend in Greece. Why?

   (b) Is it correct to say simply that the Sun rises in the east and sets in the west?

   (c) Why does not every place on the Earth have 12 hours of daylight followed by 12 hours of darkness?

5. (a) Why does Polaris seem to remain almost motionless in the sky, with all other celestial bodies apparently moving round it?

   (b) What view of Polaris would you have from (i) the north pole of the Earth, (ii) London (latitude 51½°N), (iii) the equator, (iv) New Zealand?

   (c) Explain the phenomenon of *precession*.

6. (a) What is meant by saying that a star is *circumpolar*?

   (b) What exactly is the *Zodiac*?

   (c) How would the apparent movements of the bodies in the sky be affected if the Earth rotated on its axis once in 20 hours, and took 400 days to complete one journey round the Sun?

Practical work

1. Make a model to show why the Earth's seasons occur.
2. Make a model to show why some stars are circumpolar as seen from Britain, while others are not.
3. Use a fixed camera to photograph star trails (see Fig. 158), first in the region of the celestial pole, and then in a region well away from the pole. How do the trails differ? And how can you use photographs of this kind to show that Polaris is not exactly at the polar point?

5

The Celestial Sphere

This chapter contains most of the mathematics – but please do not be alarmed! There is nothing difficult about it, and we promise not to use any complicated mathematical formulae.

Ancient peoples believed the sky to be solid. For the moment, let us suppose that they were right, and that there is in fact a transparent *celestial sphere* round the Earth, whose centre is the same as that of the Earth itself. We can then suppose that the stars are fixed on to it, and that the whole sphere revolves round the Earth once a day.*

Fig. 26 shows the celestial sphere. The north celestial pole, marked approximately by Polaris, is simply the point on the celestial sphere which lies in the direction indicated by the Earth's axis; the same applies to the south celestial pole. Next, suppose that the Earth's equator is projected on to the celestial sphere; this will cut the sky into two hemispheres, and will give us the position of the celestial equator.

We have noted that on the surface of the Earth, latitude is defined as the angular distance north or south of the equator; the polar zones lie within 23½° of the poles (that is to say, the Arctic circle is north of latitude 66½°N and its equivalent south of latitude 66½°S) while the tropics are marked by the parallels

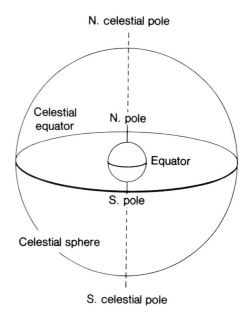

26. The celestial sphere. The celestial poles indicate the direction of the Earth's axis. The celestial equator is the projection of the Earth's equator on the celestial sphere.

of latitude 23½°N (Tropic of Cancer) and 23½°S (Tropic of Capricorn). In the same way, we can measure angular distances in the sky north or south of the celestial equator, but instead of calling it latitude we call it *declination*. Consider the bright star Betelgeux, in the constellation of Orion.* Its declination is 7°24'N, or +7°24'; therefore it is in the northern hemisphere of the sky, as shown in Fig. 27.

---

* For this chapter we can safely ignore the individual or *proper* motions of the stars, and assume that the constellation patterns are permanent. After all, the proper motions are so slight that no changes can be detected with the naked eye over several centuries – or, in most cases, over many hundreds of centuries.

*Like many star-names, this one can be spelled in several different ways; Betelgeuse and Betelgeuze are other variants.

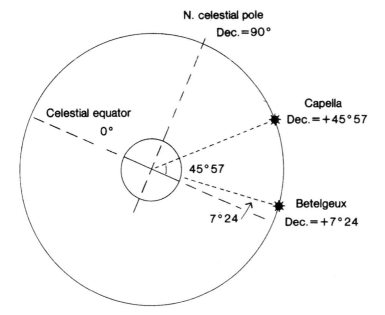

27. Declination is a star's angular distance from the celestial equator. Thus Betelgeux has dec. +7°24'; Capella, +45°57'; the north celestial pole, +90°; the equator, 0°.

Obviously, the declination of the north celestial pole is +90° and that of the south celestial pole –90°, while the declination of the celestial equator is 0° (just as the latitude of the terrestrial equator is 0°). It is easy to see why the altitude of the polar point is equal to the observer's latitude. Stand at the Earth's north pole, and the celestial pole will be straight above your head, so that its altitude will be 90°.

We can also find out which stars will be circumpolar and which will not, so let us assume that we are observing from latitude 51°N, somewhere in the London area. This means that Polaris will be 51° above the horizon (if we neglect its slight displacement from the true celestial pole, as we propose to do for the rest of this chapter). It follows that the distance between Polaris and the overhead point or *zenith distance* is 90 – 51 = 39°. To an observer in the Earth's northern hemisphere, a star is at its lowest point in the sky when it lies due north; any star which is 'below' the pole by the amount of one's latitude will just scrape the northern horizon at its lowest point. If it

is nearer the pole than that, it will never set, and will be circumpolar. As declination is measured from the equator toward the pole, we can calculate the limiting declination for a circumpolar star by subtracting our latitude from 90°, which gives us the angle *downward* from the pole. From latitude 51°N, then, a star will be circumpolar if its declination is 90 – 51 = 39°N, or greater. The brilliant yellow star Capella, with its declination of +45°57', is circumpolar from London; Betelgeux, at declination +7°24', is not.

Similarly, a star which lies at any declination *south* of –39° will never rise from London. This is why we in Britain can never see brilliant constellations such as the Southern Cross, whose declination is around –63°.

To sum up: to an observer at latitude 51°N, a star with declination greater than +39° will never set, while a star with declination below –39° will never rise. Similar calculations can be made for any other latitude. From Lerwick in the Shetland Islands, where the latitude is just over 60°N, the limiting declination

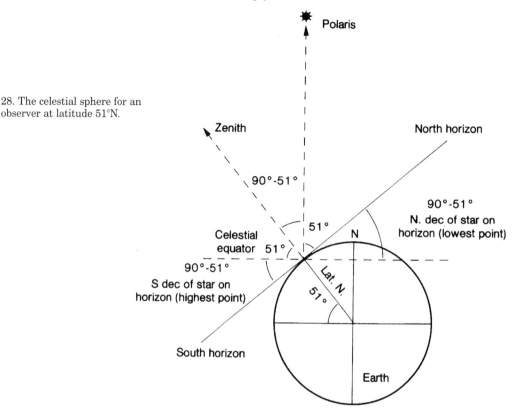

28. The celestial sphere for an observer at latitude 51°N.

will be 90 – 60 = 30°, so that a star whose declination is greater than +30° will be circumpolar. Castor, one of the two famous 'twins' (declination +32°) is circumpolar from Lerwick but not from London. If this is not perfectly clear, Fig. 28 should help.

So far, so good; we have found out how to give the celestial equivalent of a star's latitude. What we must do next is to see how we can define the 'sky equivalent' of longitude, which is known as *right ascension* (RA). This is not quite so straightforward, because we need a reference point – a kind of celestial Greenwich to mark the zero meridian – and we must begin by saying something more about the yearly motion of the Sun against the stars.

We have already noted that the Sun goes right round the sky once a year, covering 360° in just over 365 days. The

apparent yearly path of the Sun among the stars is called the *ecliptic*, and it passes through the twelve constellations of the Zodiac. Because the Earth's equator is tilted to the orbital plane by 23½°, the angle between the ecliptic and the celestial equator is also 23½°. Each year, the Sun crosses the equator twice. On or about 22 March – the date is not absolutely constant, owing to the vagaries of our calendar – the Sun reaches the equator, travelling from south to north; when it lies exactly on the equator, so that its declination is 0°, it is said to have reached the *vernal equinox* or First Point of Aries. It then spends six months in the northern hemisphere of the sky. About 22 September it reaches the equator again, this time travelling from north to south; it has reached the *autumnal equinox*, and for the following six months it lies in the southern hemisphere, so that the Earth's

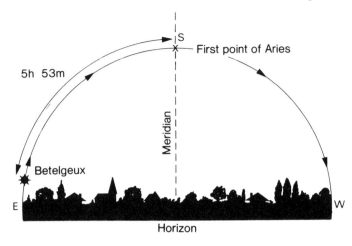

Horizon

29. The right ascension (RA) of an object is the time that elapses between the culmination of the First Point of Aries, and the culmination of the object. Here the First Point is shown on the observer's meridian, i.e. due south (assuming that the observer is in the northern hemisphere of the Earth). Betelgeux is rising. It reaches the meridian 5h 53m later, and so its RA is 5h 53m.

south pole has constant daylight and the north pole has no daylight at all.

For the moment we can forget about the September crossing, and concentrate upon the First Point of Aries. This is where the ecliptic crosses the equator; it is called the First Point of Aries because in ancient times it lay in the constellation of Aries, the Ram. By now, precession has shifted it out of Aries into the adjacent constellation of Pisces, but the name has never been altered, presumably because everybody has become so used to it.

Though right ascension is reckoned from the First Point of Aries, it is not usually given in degrees; for many purposes it is easier to use hours, minutes and seconds of time. This may sound confusing at first, but it is not really so, as a little explanation will show.

To an observer in the Earth's northern hemisphere, a body on the celestial equator will rise due east, reach its highest point or *culmination* when it is due south, and set due west. If the body lies in the southern hemisphere of the sky, it will rise south of east and set south of west, but it will still culminate when due south. Obviously, the First Point of Aries – which, by definition, lies exactly on the equator – will culminate once a day. (Remember that to an astronomer, a

day is 24 hours, not 12.) The right ascension of a star is *the sidereal time which elapses between the culmination of the First Point of Aries and the culmination of the star.* Betelgeux in Orion culminates 5 hours 53 minutes after the First Point of Aries has done so; therefore its RA is 5h 53m (see Fig. 29). Apart from the slight shifts due to precession, the right ascensions and declinations of the stars do not change; those of the Sun, Moon and planets alter constantly, because these bodies are close to us, and wander about along the Zodiac.

The great circle in the sky which passes through the pole and the zenith is called the observer's *meridian*; it is easy to see that to an observer in the northern hemisphere, the meridian will cut the horizon at the exact north and south points, so that when the First Point of Aries is at culmination, due south, it is also in *transit* on the meridian. The same is true of any other object,* but a circumpolar star will transit the meridian twice a day, once when it is 'above' the celestial pole (that is to say, south of the

* There is no bright star close to the First Point of Aries. Incidentally, precession causes the First Point to move westward by 30 seconds of arc each year.

33

polar point) and once when it is 'below' (due north, again assuming that we are observing from the Earth's northern hemisphere). These crossings are known as the *upper transit* and the *lower transit* respectively. With a non-circumpolar star, lower transit will take place when the star is below the observer's horizon. For the rest of this chapter, the unqualified term 'transit' is taken to mean 'upper transit'.

The two co-ordinates of right ascension and declination fix the star's actual position on the celestial sphere. To give the height and direction of the star as seen by the observer, we use *altitude* and *azimuth*. Altitude is, of course, the height above the horizon, reaching 90° at the zenith. Azimuth indicates direction, usually reckoned from the north point, either to 360° through east, south and west, or to 180° east and west. It is hardly necessary to add that a star's altitude and azimuth are changing all the time.

Before going any further, we must say something about time. This again is not entirely straightforward, because there are different sorts of time. The type which concerns us at the moment is *sidereal* (star) time, which is based on the true rotation period of the Earth. The Greenwich sidereal time (GST) is 0 hours when the First Point of Aries is exactly on the meridian – that is to say, at culmination – and the interval between successive culminations is 23 hours 56 minutes 4 seconds, a period known as the *sidereal day*. This is approximately four minutes shorter than the civil day of 24 hours, and the difference must always be taken into account when making calculations. The *local sidereal time* (LST) at any place is also 0 hours when the First Point of Aries is at culmination as seen from that place.

This brings us back to right ascension, because the local sidereal time must always be equal to the right ascension of a star which lies exactly on the meridian. Suppose, for example, that as seen from Plymouth the star Betelgeux is exactly on the meridian – that is to say, due south; what is the local sidereal time? We know the RA of Betelgeux; it is 5h 53m. Therefore, the local sidereal time at Plymouth at this moment is also 5h 53m. Before the development of the amazingly accurate clocks of today, the best way to check time was to observe the moment when a star of known RA passed in transit across the meridian. This was done by means of a *transit instrument*, which was simply a telescope fixed so that it always pointed to the meridian; it could move 'up and down' in altitude, but not 'from side to side' in azimuth. The observer measured the moment when the star crossed the centre of the telescopic field; since he knew the RA of the star, he also knew the time at that instant. Transit instruments are still in use at various observatories. One of them, the Swedish instrument on La Palma, is entirely automatic; switch it on, and it will take transits all by itself!

The next term we need to know is *hour circle*, which is the great circle on the celestial sphere passing through both poles and the star concerned. The angle between the celestial meridian and the star's hour circle, measured in degrees westward from the meridian, is called the *hour angle*. Think about this, and you will realise that if you want to find the local hour angle of a star (or any other body), all you need do is to subtract its right ascension from the local sidereal time. Thus, if the local sidereal time at Cardiff is 8h 55m, what is the hour angle of Betelgeux? We take the RA of Betelgeux (5h 53m) away from the local sidereal time (8h 55m), and we obtain 3h 2m, which is the local hour angle of Betelgeux at that moment. In other words, Betelgeux crossed the Cardiff meridian 3h 2m ago.

The reason that sidereal time differs from everyday time is that the Sun is moving eastward against the stars at roughly a degree a day, or 15 minutes of arc per hour. What we call Greenwich Mean Time (GMT) or Universal Time (UT) is the local time at Greenwich reckoned according to the Sun – but to complicate matters still further, we have to use not the real Sun, which travels across the sky at a variable rate, but an imaginary *mean sun*, which moves along the celestial equator at a constant rate, completing one journey per year. (This is because the Earth's orbital velocity is not constant, but is greatest near perihelion, following Kepler's Laws.) We will meet this mean sun again later, when we come to consider sundials. For the moment, it is good enough to say that as seen from Greenwich, the Sun is on the meridian at 12 hours GMT.*

If we know the local sidereal time at any observing site, we can find the Greenwich sidereal time easily enough. The Earth rotates by 1° in 4 minutes, reckoning by civil time, and so all we have to do is to allow for the longitude difference. Suppose that at Manchester, latitude 53°14'N, longitude 2°18'.4W, the star Rigel, which as an RA of 5h 12m and a declination of –8°14', is on the meridian at 20h 0m; what is the

Greenwich sidereal time? We know the local sidereal time at Manchester, because it must be equal to the RA of Rigel: that is to say, 5h 12m. Manchester is 2°18'.4 west of Greenwich. Multiplying by 4, we work out a quick sum:

$$2°18'.4$$
$$4$$
$$\overline{9°13'.6}$$

and so the difference is 9 minutes 13.6 seconds. We have to add this to the local sidereal time, because Manchester lies west of Greenwich. 5h 12m + 9m 13s.6 = 5h 21m 13s.6, which is therefore the Greenwich sidereal time at that moment.

To complete the picture, let us work out Rigel's height above the Manchester horizon at the moment of culmination. Take the latitude of Manchester away from 90°, and this gives us 36°46'. From this we must subtract Rigel's declination, because Rigel lies in the southern hemisphere of the sky. We arrive at a figure of 28°32', which is the maximum altitude of Rigel as seen from Manchester. If Rigel had been in the northern hemisphere of the sky, we would have had to add its declination to 36°46'.

The best way to explain some of the more complicated procedures is to give some worked examples. Check them through, and you will soon see how to solve other problems of the same kind.

* In summer, our civil clocks are put forward one hour to give British Summer Time; to convert BST into GMT we have to subtract one hour, but astronomy has no patience with artificial manoeuvres such as this!

Some worked examples

1. (a) What is the altitude of Polaris as seen from Rome, latitude 42°N? (You may assume that Polaris lies exactly at the north celestial pole.)
   (b) The star Alkaid, in Ursa Major, has a declination of +49°. Is it circumpolar as seen from Athens, latitude 38°N?
   (c) An observer at latitude 57°N finds that the star Capella has an altitude of 13° when it is at lower transit, i.e. directly under the pole, due north. Calculate the declination of Capella.

Answers

(a) The altitude of the celestial pole is always equal to the observer's latitude. From Rome, therefore, it will be 42°.

(b) To find the limit for a circumpolar star, we must take the observer's latitude away from 90°. 90 – 38 = 52. Therefore from Athens, a star will be circumpolar only if it is north of declination +52°. Alkaid is only 49°N, so that from Athens it is not circumpolar.

(c) We know that declination = altitude + zenith distance. In this case the zenith distance is 33° (since 90 – 57 = 33). Adding this to the altitude of Capella at lower transit gives us the star's declination. 13° + 33° = 46°, which is therefore the declination of Capella.

2. Vega (RA 18h 35m, dec. +38°44') was observed at upper transit of the meridian at Edinburgh (lat. 55°57'N, long. 3°11'W) on 6 July 1968, at 23h 48m GMT.* Find:

(a) The altitude of Vega at this moment, as seen from Edinburgh.

(b) The Greenwich Sidereal Time at this moment.

(c) The GMT when Vega will be in upper transit from Edinburgh on 13 July 1968.

Answers

(a) Find the co-latitude of Edinburgh, i.e. the result of subtracting the latitude from 90° (also equal to the meridian altitude of the celestial equator, i.e. a star with declination 0°):

$$90° – 55°57' = 34°3'.$$

To this we must add Vega's declination, because Vega lies in the northern hemisphere of the sky. 34°3' + 38°44' = 72°47', which is the required altitude of Vega at upper transit as seen from Edinburgh.

(b) The Local Sidereal Time at Edinburgh at the time of transit must be equal to Vega's RA: that is to say, 18h 35m. We must allow for the longitude of Edinburgh, which is 3°11'W. Multiplying 3°11' by 4, we obtain 12 minutes 44 seconds. 18h 35m + 12m 44s = 18h 47m 44s, which must therefore be the Greenwich Sidereal Time at this moment.

(c) On 6 July the GMT of the transit was 23h 48m. Seven days later, on 13 July, the transit will be earlier. The rate of difference is approximately 4 minutes per day, and 7 x 4 = 28:

$$23h 48m – 28m = 23h 20m.$$

If we want better accuracy, the daily difference can be taken as 3m 57s instead of exactly 4 minutes. This makes the total difference 27m 39s instead of 28m, and our answer works out to 23h 20m 21s.

3. At 0 hours GMT on 20 June, the Greenwich Sidereal Time was 17h 51m. Find:

(a) The GMT on this date when the star Deneb, RA 20h 39m 42s, dec. +45°6', was at upper transit on the meridian at the Lowell Observatory, Arizona, lat. 35°11'N, long. 111°45'W.

(b) The altitude of Deneb at this transit.

(c) Whether Deneb is circumpolar as seen from the Lowell Observatory.

*For all practical purposes we can say that Greenwich Mean Time (GMT) and Universal Time (UT) are the same. Use whichever you like for our present purpose!

Answers

(a) We know the Local Sidereal Time at Lowell when Deneb was on the meridian, because it must be equal to the RA of Deneb (20h 39m 42s). Therefore, we can find the Greenwich Sidereal Time by allowing for the latitude of Lowell, 111°45'W. We divide this figure by 15 and obtain our answer in hours, minutes and seconds of time:

$$111°45' \div 15 = 7\text{h } 27\text{m } 0\text{s}.$$

We can now proceed. Since we are working in fairly good approximations, we can find the GMT of the transit at Greenwich, which we can do by subtracting the Greenwich Sidereal Time, at 0 hours GMT, from the RA of Deneb:

$$20\text{h } 39\text{m } 42\text{s} - 17\text{h } 51\text{m } 0\text{s} = 2\text{h } 48\text{m } 42\text{s}$$

which is the GMT of the transit at Greenwich. Now we allow for the longitude of Lowell:

$$2\text{h } 48\text{m } 42\text{s} + 7\text{h } 27\text{m } 0\text{s} = 10\text{h } 15\text{m } 42\text{s}$$

(we add, because Lowell is west of Greenwich).

Again in good approximation, this gives us the GMT of the transit as seen from Lowell, but we really ought to make another correction, because we have used Greenwich Sidereal Time at 0 hours GMT, whereas we ought to have used the Greenwich Sidereal Time at 10h 15m 42s GMT. As we have seen, the correction is approximately 4 minutes per day, which we can take as being 10 seconds per hour. In 10h 15m 42s, this mounts up to 1m 43s, which we should subtract – giving us a final answer of 10h 13m 59s.

(b) As before, take the latitude of Lowell away from 90°:

$$90° - 35°11' = 54°49'.$$

Add the declination of Deneb, which is north of the celestial equator:

$$54°49' + 45°6' = 99°55'.$$

But nothing can have an altitude of greater than 90°. Deneb is 9°55' 'over the top', so to speak, so that its altitude above the horizon is 90° – 9°55' = 80°5'; it is bearing north, not south.

(c) Again as before, take the latitude of Lowell away from 90°, giving 54°49'. This means that any star with declination greater than 54°49' will be circumpolar. However, Deneb is not so far north as this; its declination is only +45°6', so that it will not be circumpolar as seen from Lowell.

4. On 10 December, 1968, Rigel (RA 5h 12m 6s, dec. –8°14') was at upper transit on the meridian at Manchester, lat. 53°14'.2N, long. 2°18'.4W, at 0h 08m.2 GMT. At 2h 32m.6 GMT on the same date, the star Procyon was on transit of the same meridian, south of the zenith, at an altitude of 42°8'. Find:
(a) The Greenwich Sidereal Time of the transit of Rigel.
(b) The Greenwich Sidereal Time of the transit of Procyon.
(c) The RA and declination of Procyon.

Answers

(a) The Local Sidereal Time at Manchester was, of course, equal to the RA of Rigel, i.e. 5h 12m 6s, so all we need do is to allow for the longitude of Manchester. This is 2°18′.4. Multiplying by 4, we arrive at a correction of 9m 13s.6 which for calculations of this kind we are entitled to round off to 9m 14s. We have to add it to the Local Sidereal Time, because Manchester lies west of Greenwich:

$$5h\ 12m\ 6s + 9m\ 14s = 5h\ 21m\ 20s,$$

which is the Greenwich Sidereal Time of the transit of Rigel.

(b) Rigel passed in transit at 0h 8m.2, or 0h 8m 12s GMT. Procyon was at transit at 2h 32m.6, or 2h 32m 36s – that is to say, 2h 24m 24s later reckoning by GMT. But we must add another 24 seconds, because of the difference between the rates of solar and sidereal time, which amounts to 10 seconds per hour. So the difference between the transits of Rigel and Procyon comes out at 2h 24m 8s. Now:

Greenwich Sidereal Time at 0h 8m 12s = 5h 21m 20s
Difference in sidereal time   = 2h 24m 48s
                                     7h 46m 8s

which must be the Greenwich Sidereal Time of the transit of Procyon.

(c) The Greenwich Sidereal Time of the transit of Procyon was 7h 46m 8s, as we have just found. Manchester's longitude is 2°18′.4W, which, when converted to time, amounts to 9m 14s. We must subtract, because Manchester is west of Greenwich:

$$7h\ 46m\ 8s - 9m\ 14s = 7h\ 36m\ 54s$$

which was the Local Sidereal Time of the transit of Procyon, and must also, by definition, be the right ascension of Procyon.

We next need the co-latitude of Manchester, which we find by taking Manchester's latitude away from 90°:

$$90° - 53°14′.2 = 36°45′.8,\ or\ 36°45′48″.$$

This must be subtracted from the given altitude of Procyon at transit:

$$42°8′ - 36°45′48″ = +5°22′12″$$

which is the declination of Procyon.

Beware of one possible confusion. Because of precession, the right ascensions and declinations of stars alter slightly from year to year – so that if you check some of the positions given here with the values given in catalogues of different date, you will find slight discrepancies. Fortunately these make no difference to the calculations themselves.

Questions

1. Draw a diagram to show the celestial sphere, putting in the celestial poles, the celestial equator, and the poles and equator of the Earth.

2. (a) Define the following terms: ecliptic, hour angle, great circle, culmination, First Point of Aries, zenith distance.

    (b) What is meant by the right ascension and declination of a star? How does this differ from altitude and azimuth?

    (c) Why does a clock which keeps GMT run at a slower rate than an observatory clock which keeps sidereal time? What is the difference per hour?

3. At 0 hours GMT on 6 April 1967, the Greenwich Sidereal Time was 12h 54m. Find:

    (a) The local hour angle of Betelgeux, RA 5h 53m, dec. +7°24', for an observer at Broadstairs, lat. 51°21'N, long. 1°26'E, at 18h GMT.

    (b) The altitude of Betelgeux when it is on the meridian at Broadstairs.

    (c) Whether Vega, dec. +38°44', will be circumpolar as seen from Broadstairs.

4. An observer at Bristol, lat. 51°27'N, long. 2°33'W, finds that the star Altair, RA 19h 48m, dec. +8°44', is at transit at 22h 20m GMT. Find:

    (a) The altitude of Altair at upper transit.

    (b) The Greenwich Sidereal Time of the transit at Bristol.

    (c) The GMT of the transit of Altair as seen from Bristol one week later.

5. At 0 hours on 1 March 1968, the Greenwich Sidereal Time was 10h 35m 26s. Find:

    (a) The GMT on the same date when the star Mizar, RA 13h 21m 54s, dec. +55°11', was at upper transit at the McGill Observatory, Montreal, lat. 45°30'20"N, long. 73°34'42"W.

    (b) The altitude of Mizar at this transit.

    (c) The local hour angle of Mizar as seen from McGill Observatory at 20h 0m GMT on the same date.

*

The answers to questions 3, 4 and 5 are given in the Appendix, but don't look at them until you have worked through the calculations and are quite sure that you understand them!

6

The Measurement of Time

Astronomical measurement is the basis of all timekeeping. Indeed, this was one of the main reasons why the ancients were so concerned about it. If, for instance, you happened to live in Ancient Egypt, it was an excellent thing to know just when the Nile might be expected to flood – as happened each year at about the same time.

Calendars were drawn up a long time ago. Of course they were based on the Earth's period of revolution round the Sun; the fact that the ancients believed the Sun to go round the Earth instead of vice versa made no difference so far as their calendars were concerned. Unfortunately all calendars, even the modern one, are bound to be somewhat awkward, because the Earth does not go round the Sun in an exact number of days. A complete revolution takes not 365 days, but 365¼. To have a calendar taking account of an extra quarter-day would be hopelessly clumsy, but neither can we simply forget about the error, because the seasons would become out of step – and after a while we would find that, for example, Christmas in Britain would occur in the summer, and winter in June.

What we do is to add an extra day every 4 years, to compensate for the error of a quarter of a day per year. The additional day is tacked on to February, the shortest of the twelve months; the year has 366 days instead of 365, and is called a leap year. To decide whether a year is a leap year or not, divide by 4; if there is no remainder, February will have 29 days instead of 28. Thus 1996 was a leap year (1996 ÷ 4 = 499; remainder 0) but 1997 was not.

This system was worked out on the orders of no less a person than Julius Caesar, but it was still not quite accurate enough. In the modern Gregorian calendar, named after Pope Gregory XIII, a century year (1700, 1800, 1900, etc) is a leap year only if it is exactly divisible by 400. Thus 1900 was not a leap year, but 2000 was. There is still a minor error; but as it will take 3,300 years for this error to mount up to one day, it is small enough to be disregarded.

Just as the 'year' is related to the Earth's period of revolution round the Sun, so the 'day' is linked with the Earth's axial rotation period. For convenience, we obviously need to take an exact number of hours for our civil day, and astronomically we use the 24-hour clock. Remember that a solar day is slightly longer than a sidereal day, because the Sun is moving eastward against the stars. It shifts about 1° per day against the starry background, because it completes the full circuit (360°) in just over 365 days.

Nowadays, anyone who wants to know the time need do no more than look at his wrist-watch. More accurately he can tune in to a radio time signal, or dial the Speaking Clock on the telephone. None of these convenient methods could be used in ancient times, and the best clocks were sundials, which are neither accurate nor easy to use, but which are a great deal better than nothing at all. Later came water-clocks or 'clepsydrae', which depended upon the rate at which water flowed out of a graduated container, but for the moment let us concentrate upon telling the time by the Sun.

The ancestor of the sundial was the

shadow-clock, which consisted of nothing more elaborate than a pole stuck upright in the ground. As the Sun moved across the sky, the shadow of the pole would be thrown on to the ground, and a graduated dial would give some idea of time-intervals. The trouble about this was that the rate of the shadow movement varies with the declination of the Sun, so that a shadow-clock accurately graduated for one season would be inaccurate at another.

The remedy worked out by the Arabs was to put the pole, or *style*, at such an angle that it was parallel with the Earth's axis. This turned the shadow-clock into a true sundial. The style will always stay parallel with the axis of the Earth – in other words, it will point to the pole of the sky – so that the shadow cast by the Sun will move at a steady rate, and all we need do is to calibrate the dial into divisions of 15° each. Each division will then represent one hour of time (Fig. 30)

To make a sundial of this kind, mark out your dial, put in the style vertically to the dial, and then mount the style so that

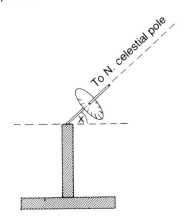

31. Simple sundial. The angle x = the observer's latitude. The trouble is that for half the year the shadow will fall on the underside of the dial, and the instrument is useless.

it points to the celestial pole, i.e. is set at an angle equal to your own latitude. This arrangement will work quite well (Fig. 31), subject to some minor corrections to be noted below. Unfortunately, there is one major snag. For half the year the Sun's rays will be lower than the dial, so that the shadow will fall on the underneath and the instrument will be useless. In Britain, this will happen when the Sun is south of the celestial equator, i.e. between late September and late March.

To make a sundial which will work throughout the year, we must keep the style at an angle equal to the latitude of the observer, but the dial must be horizontal (Fig. 32). The angular movement of the Sun's shadow will no longer be steady, but it will not change from one season to another, so that we can allow for it by calibrating the dial suitably. There are mathematical formulae for doing this, but there is also a simple graphical method which will serve, and we have given it at the end of this chapter. As soon as you have the style mounted, the dial calibrated and the whole instrument set up, you will be able to use your sundial – clouds permitting!

Unfortunately you cannot simply go to

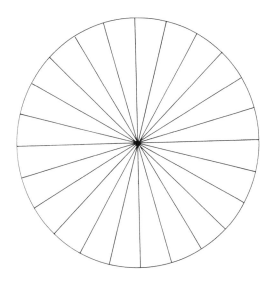

30. The dial divided into sections of 15° each.

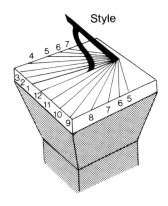

Style

32. Sundial; this will work throughout the year.

the sundial, note the position of the shadow, and read off the correct time. First, the sundial will always show local time, not GMT. Secondly, we must allow for what we call the *equation of time*.

Because the Earth's path round the Sun is not circular, the orbital velocity changes; the Earth moves quickest when it is at perihelion, or closest to the Sun (that is to say, in northern winter). This means that the Sun's speed against the background of stars will also be variable. To allow for this in everyday timekeeping would be most awkward, and so astronomers use a completely fictitious *mean sun*, which is assumed to travel along the celestial equator at a constant rate, completing one circuit in the same time that the real Sun takes to go round the ecliptic. We may compare the two suns with two runners, who go round a circular track and dead-heat at the winning-post – but one runner (the mean sun) has kept up a constant speed throughout, while the other (the real Sun) has sometimes spurted ahead, sometimes lagged behind. The difference in position between the mean sun and the real Sun causes a difference between apparent solar time and mean solar time. This difference is called the Equation of Time, and may rise to over 16 minutes. Luckily it is the same for every year, and on four dates it becomes zero. The values for various

dates are given in the table. (We know that this explanation is incomplete, but it is good enough for our present purpose.)

A sundial shows local time, not Greenwich time, so that to convert the sundial reading to GMT we have to allow for our longitude as well as for the equation of time. In Britain the difference is never great, because our islands are small; even so, local noon in Land's End is over twenty minutes later than Greenwich noon.

There is no point in allowing for this in civil life, but countries with a larger area have to divide themselves up into various time zones. Scientifically, GMT (or UT if you like) is always used; it is simply the local time at Greenwich reckoned according to the mean sun. But for everyday use, time zones are essential. Thus Mexico is 6 hours behind GMT, Iceland 1 hour behind, New Zealand 11h 30m ahead, and so on. In North America there are five time zones – Pacific, Mountain, Central, Eastern and Atlantic – which are respectively 8, 7, 6, 5 and 4 hours behind GMT.

The Equation of Time

A minus sign (–) indicates that the clock is ahead of the Sun; a plus (+) that it is behind the Sun. Clock time = (sundial time) – (equation of time). So if the sundial time is 10.30 and the equation of time is –7 minutes, then clock time = 10.30 – (–7) = 10.37.

Example

Suppose that we are observing from Teignmouth in Devon (long. 3°30'W) and find that on 1 January the sundial shows 1.23 p.m. What is the GMT?

From the table (opposite), the equation of time is –3 minutes; this brings us to 1.26 p.m., or 13.26 on the 24-hour clock. The longitude of Teignmouth is 3°30'W. Multiplying by 4, we obtain a correction of

6. The Measurement of Time

		min.			min.			min.
Jan.	1	−3	Apr.	25	+2	Sept.	2	0
	9	−7	May	7	+3½		14	+4
	18	−10		15	+3¾		28	+9
	24	−12		28	+4	Oct.	7	+12
Feb.	4	−14	June	10	+1		20	+15
	12	−14½		15	0		27	+16
	24	−13½		29	−3	Nov.	4	+16½
Mar.	4	−12	July	10	−5		18	+15
	16	−9		27	−6⅓		29	+12
	26	−6	Aug.	13	−5	Dec.	4	+10
Apr.	5	−3		29	−1		13	+6
	16	0					26	0
							30	−2

14 minutes, which must be added, because local time at Teignmouth is earlier than local time at Greenwich. 13.26 + 14 minutes = 13.40, which is the GMT.

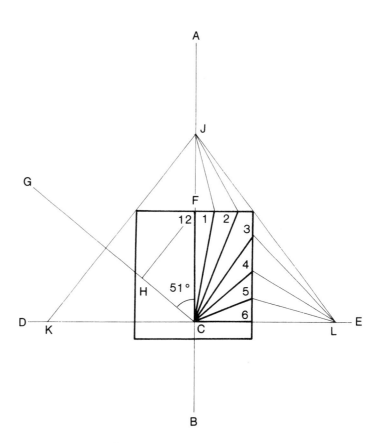

33. Making a sundial (see p. 44 for explanation).

43

Making the dial (Fig. 33)

Divide a square, of any size, into two by the line AB. Select a point C for the position of the style. Draw DCE, at right angles to AB, which will be our 6 o'clock line.

Draw CG, making angle ACG equal to your latitude; this dial has been drawn for latitude 51°. From F, the top point of intersection of AB with the square (our 12 o'clock point), drop a perpendicular FH on to CG. On AB, mark off FJ, making FJ = FH.

From J, construct angles of 15° each, obtaining the 1 o'clock, 2 o'clock points and so on. When these lines no longer hit the top of the square, draw JK and JL through the corners of the square; K and L will fall along DE. From K and L construct 15° angles as before, so obtaining the rest of the hour-lines on the side of the dial. Connect all the hour-lines up to the point C – and the dial is complete. In the diagram not all the lines have been drawn in, as this would make it crowded, but the idea is clear enough.

Questions

1. Why is it necessary to have a leap year every four years? Which of the following are leap years: 1900, 1934, 1977, 1988, 1989, 1990, 1995, 2000, 2002?

2. Explain how a sundial may be used to measure solar time. How does this time differ from mean time – and why?

3. (a) Explain the difference between a shadow-clock and a true sundial.
   (b) Draw the dial for a sundial to be set up by an observer at Redruth, lat. 50°15'N, long. 5°13'W.

4. (a) Construct a graph, using the values given in the table on p. 42, from which the equation of time can be read off for any date in the year.
   (b) On 20 December our observer at Redruth finds that his sundial reads 10.50 a.m. What is the GMT at this moment?

5. Explain the meaning of Zone Time. Why does North America have five different zones, Britain only one?

7

The Electromagnetic Spectrum

Until fairly recently, astronomers had to depend almost entirely upon studying the light received from bodies in the sky. This is no longer true – what we often call 'invisible astronomy' has now come very much to the fore – but before going into detail, we must say something about radiation in general.

Light may be regarded as a wave motion, and the colour of the light depends upon its wavelength, i.e. the distance between one wave-crest and the next (Fig. 34). The velocity of light is absolutely constant. In round numbers it is 300,000 km per second. Scientifically, this is usually written as 3×10^5 km sec^{-1}, because there is a special way of giving very large or very small quantities. The system is based upon the number ten. 10^1, or 10 to the power of 1, is of course simply 10; 10^2, 10 to the power of 2, is $10 \times 10 = 100$; 10^3 is 1000, and so on. 10^5 is therefore 100,000 so that $3 \times 10^5 = 300,000$. It is a convenient form of mathematical shorthand, and can also be used for values below 1. Thus 10^{-1} = 1/10; 10^{-2} = 1/100, and so on.

Light has very short wavelength. The basic idea is simple enough – as you can see if you throw a stone into a calm pond – but with light we are dealing with tiny fractions of a millimetre, and ordinary units of measurement become clumsy. We have to use different units, and three are of special importance:

The *micron* (μ), which is one-millionth of a metre (10^{-6}m).

The *nanometre* (nm), which is one thousand-millionth of a metre (10^{-9}m).

The *Ångström*, which is one ten-thousand millionth of a metre (10^{-10}m).* One Ångström is equal to one hundred-millionth of a centimetre. Mentally divide a line one centimetre long into a hundred million parts, and you may have some idea of how small an Ångström really is.

Visible light forms only a very small part of the total range of wavelengths, or *electromagnetic spectrum*. If the wavelength is longer than that of red light, or shorter than that of violet light, it does not affect our eyes, and we have to detect it in different ways. The main regions of the electromagnetic spectrum are:

Below 0.01 nm: gamma-rays.
0.01 nm to 10 nm: X-rays.
10 nm to 400 nm: ultra-violet.
400 nm to 700 nm: visible light. (This is equivalent to 4000 Å to 7000 Å, but we are giving rounded-off values; the true range of visible light is more like 3900 Å to 7500 Å.)
700 nm to 1 mm: infra-red.
1 mm to 0.3 m: micro-wave.
Over 0.3 m: radio radiation.

Infra-red can be detected easily enough merely by switching on an electric fire; you will feel the infra-red, as heat, well before the bars become hot enough to glow in visible light. And this demonstrates an important principle. Generally speaking, the higher the temperature of the body

---

* The name was given in honour of the nineteenth-century Swedish physicist Anders Ångström. It was inconvenient of him to have a surname beginning with the distinctive Swedish letter Å!

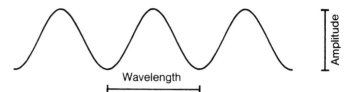

34. Wavelength.

sending out radiation, the shorter will be the wavelength of the radiation, so that cool bodies emit infra-red and radio waves, while very hot ones are strongest in gamma-rays, X-rays and ultra-violet.

The Earth's atmosphere blocks out most of the radiation in the electromagnetic spectrum; there are only two major 'windows', one (naturally) for visible light and the other for some types of radio waves. Infra-red is largely blocked by the water-vapour in our air, but this can be overcome to a large extent by setting up equipment at high altitude; as we have seen, the world's largest infra-red telescope (UKIRT) is situated on the top of Mauna Kea in Hawaii. There is also the problem that so far as infra-red is concerned, the telescope itself is 'hot', and produces its own radiation. This means that the detectors have to be kept intensely cold. However, this can be managed, and it has even been possible to make a camera that produces actual pictures in the infra-red.

Radio astronomy may be said to have been born in 1931, when an American radio engineer, Karl Jansky, was using a makeshift aerial to study 'static'. To his surprise, he found that he was picking up radio emissions from the Milky Way. His discovery caused very little excitement at the time, but after the war large radio telescopes were constructed – most people know the impressive 76-m (250-foot) 'dish' at Jodrell Bank in Cheshire, now known as the Lovell Telescope. Actually, the term 'radio telescope' is rather misleading, because one certainly cannot look through it; it is more in the nature of a powerful aerial, and by no means all

radio telescopes are 'dishes' – some of them have been said to look more like arrays of bean-poles. The radio emissions are collected and focused on to the aerial, where they are analysed by a computer. The usual end product is a trace on a graph, which may not look startling, but which can tell us a great deal. ('Radio noise' is another misleading term. The hissing and crackling from the Sun which has so often been broadcast on the radio and television is produced in the equipment, and is only one method of recording the radio waves. Sound cannot travel in a vacuum, and there are 150,000,000 km of vacuum between ourselves and the Sun.)

With other radiations the situation is even worse, because they are completely blocked by the layers in the Earth's atmosphere, and the only solution is to send our equipment above the screen. This has been possible only since the start of the Space Age in 1957; for example, X-ray astronomy had to wait until 1962. By now we have many artificial satellites made specially for the purpose. The latest X-ray satellite, Chandra, was launched in 1999.

There is one principle which has become all-important in certain branches of astronomy, and which we ought to introduce here even though we will be dealing with it more fully later on. This is the Doppler effect, named after the nineteenth-century Austrian physicist Christian Doppler. Again it is easy to give an everyday demonstration of what it means – and most people must have noticed it even without realising its significance.

Consider an ambulance or a police car which is coming towards you, sounding its horn. The note of the horn is high-pitched. As soon as the vehicle has passed by, and has started to recede, the note of the horn drops. When the vehicle is approaching, more sound-waves per second are reaching your ear than would be the case if the vehicle were standing still; this increases the *frequency*, and the wavelength of the sound seems to be shortened. With a receding vehicle, the frequency is lowered; fewer sound-waves per second reach you, and the note of the horn drops.

With light there is the same basic result, this time affecting the colour. An approaching light-source looks a little more blue than it would otherwise do ('blue shift') while a receding light-source looks a little too red ('red shift'). Of course,

the actual change in colour is too small to be noticed at everyday speeds, but astronomical bodies can move at tremendous velocities, and the Doppler effect makes its presence felt, since it shows up as soon as we examine the spectrum of a moving object.

We have come a long way since astronomers were limited to studying visible light alone. They were then very much in the position of a pianist who is trying to play a piano which has no notes apart from those of the middle octave. Now that we can extend our range into the other parts of the electromagnetic spectrum (Fig. 35), we can learn much more – even if we have had to develop entirely new instruments, and to make use of our new ability to send satellites and space-probes high above the Earth's atmospheric screen.

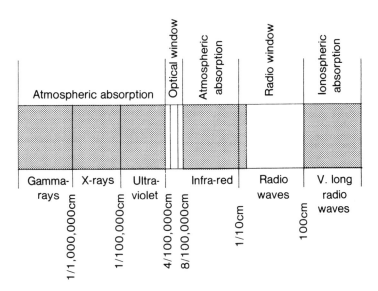

35. The electromagnetic spectrum.

Questions

1. (a) What is meant by wavelength?

(b) Which is the longer – a nanometre or an Ångström?

(c) What is the difference between wavelength and frequency?

2. (a) Why does the detector of an infra-red telescope have to be kept very cold?

(b) Why is UKIRT, the United Kingdom Infra-Red Telescope, set up on the top of Mauna Kea in Hawaii?

(c) Why do we have to use rockets and space-probes for X-ray astronomy?

3. Arrange the following wavelengths in order of length, beginning with the shortest: infra-red, visible light, ultra-violet, radio waves, microwaves, gamma-rays, X-rays.

4. Describe the main principle of a radio telescope such as the Lovell Telescope at Jodrell Bank.

5. Explain the principle of the Doppler effect. If a source of light shows a red shift, is it approaching or receding from the observer?

8

Telescopes

As soon as you become really interested in astronomy, you may well start to think about buying a telescope. There are traps for the unwary here; but first, let us say something about telescopes in general. There are two main types, refractors and reflectors.

The principle of the *refractor*, or refracting telescope, is very simple. The light from the object under study is passed through a specially-shaped lens known as an object-glass or objective, lettered O in Fig. 36. The rays are brought to focus at the point P, and an image of the object is produced, which is then enlarged by a second lens known as an eyepiece or ocular. The eyepiece is really a special form of magnifying-glass. If the telescope has an object-glass 3 inches (7.6 cm) in diameter, it is known as a 3-inch or 7.6-cm refractor. The distance OP, between the object-glass and the focus, is called the *focal length*; the focal length divided by the diameter of the object-glass is the *focal ratio*. Thus if our 3-inch (7.6-cm) refractor has a focal length of 36 inches (91.4 cm), its focal ratio is 36/3 = 12 (in Metric, 91.4/7.6 = 12).

Despite the efforts of our rulers to force everyone to use the Metric system, many telescopes are still listed in the old Imperial units, which are a great deal more convenient, so that for most of this chapter we think it will be best to use both. We will also 'round them off', so that, for example, we will take 6 inches as being equivalent to 15 cm instead of the more accurate 15.2 cm.

The object-glass collects the light, but the actual magnification is done by the eyepiece, which has its own focal length. The focal length of the object-glass divided by the focal length of the eyepiece gives the *magnification*; if M is the magnification, F is the focal length of the object-glass and f is the focal length of the eyepiece, then M = F/f. Suppose that with our 3-inch (7.6-cm) refractor with its focal length of 36 inches (91 cm) we use an eyepiece of focal length ½ inch (1.3 cm). The magnification will be 36/½ (91/1.3) = 72, often written as x 72. (Yes, we know that the Metric figures do not quite work out; but if you put in the more accurate values of 91.4 cm and 1.27 cm, you will find that they are correct!)

In theory, and usually in practice, eyepieces are made with a standard thread, so that any eyepiece can be used with any telescope. With any telescope, it is almost essential to have several eyepieces – for instance, one to give low magnification and wide field, suitable for observing star-fields; one with a moderate magnification, for views of the Moon, planets and objects such as double stars; and one with high magnification, for detailed views on good, clear nights. It is a general rule that a telescope will bear a magnification of x 50 per inch (2.54 cm) of aperture. For our 3-inch, f/12 refractor we might then select three eyepieces; a 1-inch (magnification 36/1 = 36): a ½-inch (36/½ = 72) and a ¼-inch (36/¼ = 144).

Suppose, however, that with the same telescope we want to experiment with an eyepiece of focal length about ⅛ of an inch, or 0.3 cm? As before, divide the focal

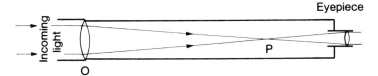

36. Principle of the refractor:
O = object-glass; P = focal point.

length of the object glass by that of the eyepiece. $36/^1/_8$ (or $91.4/0.32$) = 288. It may sound very nice to be able to use a power of 288 on the telescope; but every time an image is enlarged, it becomes fainter – and with this power on this telescope, the image would be so faint that it would be completely useless.

No; if we want to use a magnification as high as this, we must have a larger object-glass to collect the light. A 6-inch (15.2-cm) refractor of focal length 72 inches (183 cm) will have a focal ratio of f/12. On this, an eyepiece of focal length ¼ inch (0.6 cm) will give a magnification of $72/^1/_4$ = 288, which is quite acceptable; the larger object-glass will collect enough light to make the image reasonably bright with such a magnification. Of course, the focal ratio need not be 12. If the focal length of our 6-inch (15.2-cm) refractor is only 54 inches (137 cm), the focal ratio will be 54/6 = 9. This makes for greater convenience, because the tube is shorter, but to make the f/ratio too low will introduce other troubles, and one has to strike a happy mean.

Refractors are easy to handle, and they do not go out of adjustment unless they are roughly treated. Of course, they have their disadvantages. As we have noted, light is really a blend of all the colours of the rainbow, and these various colours are not bent or refracted by the same amount when they are passed through a lens. The shorter wavelengths (violet and blue) are refracted most; the longest wavelengths (red) least. The result is shown in Fig. 37, which is not to scale. The red light is brought to focus at a greater distance from the object-glass, and a bright object, such as a star, will be apparently surrounded by coloured rings which may look pretty, but are a thorough nuisance to the astronomer.

This false colour trouble can be reduced by means of what is termed an *achromatic* objective, in which there are several component lenses made of different kinds of glass and held together in their cell. (It is unwise for the telescope-owner to take them out of their cell, for examination or any other purpose, unless he is really skilled in optics. Leave well alone.) Some false colour always remains, but with a good achromatic or colour-free objective it is not really evident. If you look through your telescope and see bright, gaudy rings, you will know that the optics are of poor quality – though, of course, the fault may lie in the eyepiece rather than the main object-glass.

As almost everyone knows, an astronomical refractor will give an inverted or upside-down image. In fact,

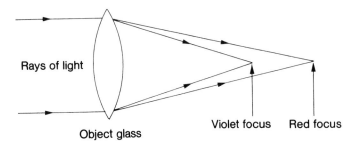

37. False colour (known technically as chromatic aberration). The drawing is not to scale.

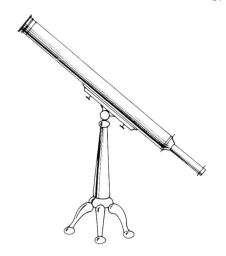

have to make use of every scrap of light available. Therefore, the correcting lenses are left out, and the image remains upside down. Astronomical drawings and photographs are often oriented with south at the top.

The higher the magnification used, the smaller will be the field of view, and so it is essential to have a telescope which is firm and easy to handle. The simplest form of mounting is the *altazimuth*, which means that the telescope can be swung freely either in altitude (up and down) or *azimuth* (to and fro). A tripod is suitable; a small, wobbly mount of the type known as the pillar and claw (Fig. 38) is to be avoided, as it will certainly be unsteady.

38. Pillar and claw mounting. I have nicknamed this the blancmange mounting, for reasons which are obvious to anyone who has tried to use it with a high magnification! (P.M.)

There is, however, an ever-present problem. As the Earth spins on its axis, all the celestial bodies move across the sky, so that the telescope has to be shifted constantly to keep the target object in view. With an altazimuth mounting there are two movements to worry about (altitude and azimuth), but things are easier if the mounting is *equatorial*. One common form of equatorial mounting, the German, is shown in Fig. 39. There is a polar axis which is aimed at the pole of the sky, so that it is parallel to the axis of the Earth; the telescope is mounted on one end of the polar axis, and there is a

any refractor will do this; but in a telescope made for terrestrial use, an extra lens-system is put into the optical train to turn the image the right way up again. However, each time a ray of light passes through a lens it becomes slightly weakened. This does not in the least matter when we are looking at a mountain-top or a ship out to sea, but it matters very much when we are studying very faint astronomical bodies, and we

39. Reflecting telescope on a German-type mounting, with its counterweight.

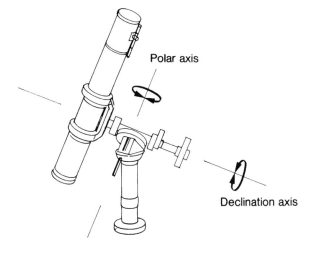

Polar axis

Declination axis

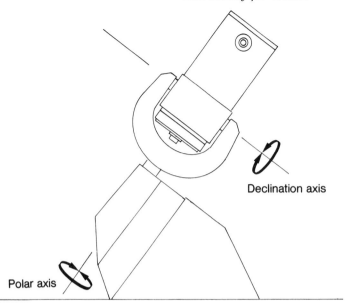

40. Fork mounting.

counterweight on the other end. When the telescope is moved round in azimuth (RA), the up-and-down motion will look after itself. Manual slow motions can be fitted, but the ideal is to add a driving mechanism, usually electrical, so that the telescope is moved round at a rate which exactly compensates for the rotation of the Earth. A driven equatorial is more or less essential for any telescopic astronomical photography.

There are various other forms of equatorial mounting – the Fork, for instance (Fig. 40) and the English (Fig. 41), in which the polar axis is mounted between pillars and the telescope is pivoted inside it – but the principles are the same. For serious observing the equatorial has overwhelming advantages, particularly if driven, though for casual 'looking around' an altazimuth is perfectly suitable.*

*With modern computers, it is now possible to drive an altazimuth simultaneously in both altitude and azimuth. The great telescopes now being built are mounted in this way, and many advanced amateurs have followed suit – but this is not something to be tackled by the newcomer.

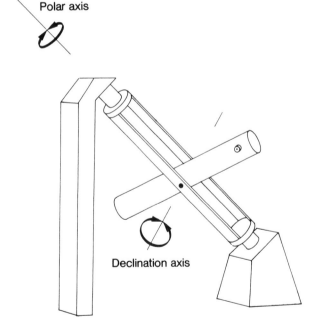

41. English mounting.

42. Observatory for a 22-cm reflector; the top part of the building rotates.

The chief trouble about refractors, from the viewpoint of the average beginner, is their price. A decent 3-inch (7.6-cm) instrument will cost something of the order of £400 if it is bought new, and larger refractors are prohibitively expensive. We will have more to say about this later.

The second type of astronomical telescope is the *reflector*, in which the light-collecting is done by using a mirror instead of an object-glass. The most common form is the Newtonian. (The name comes, of course, from Sir Isaac Newton, who built the first telescope of this type more than three hundred years ago.) Here, the light comes down an open tube and hits a curved mirror, M in Fig. 43. The light is reflected back up the tube on to a smaller flat mirror, F; the flat is placed at an angle of 45° and directs the

light-rays to the side of the tube, where an image is formed and magnified by an eyepiece as before. With a Newtonian, the observer looks into the tube instead of up it. The f/ratio depends upon the focal length of the main mirror or speculum. The mounting may be altazimuth, but here too an equatorial is much to be preferred.

Aperture for aperture, a mirror is less effective than a lens. No Newtonian reflector with a mirror less than 6 inches (15 cm) in diameter is of much use for proper observing. Everything depends, too, on the quality of the optics. If the curve of the main mirror is faulty, the telescope will give poor results – and the same is true if the flat mirror is not genuinely flat. Unfortunately, one cannot always judge by 'looking', and no beginner should buy a reflector without taking skilled advice.

The mirror has to be coated to make it really reflective. A thin layer of silver is sometimes used, but most mirrors are coated with aluminium. Periodically they will become tarnished, and will have to be re-aluminised. Against this, a reflector is much cheaper than a refractor of equal light-grasp, and there is no false colour trouble, because a mirror reflects all wavelengths equally. Obviously the flat in a Newtonian gets in the way of the incoming light to some extent, but the loss is not serious.

There are many types of reflectors. In particular there is the Cassegrain (Fig. 45), in which the secondary mirror is convex, and the light is reflected back through a hole in the main mirror. There are also more complicated instruments in

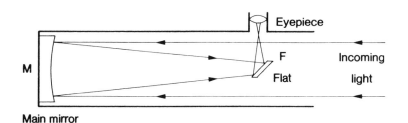

Eyepiece

F

Flat

Incoming

light

M

Main mirror

43. Principle of the Newtonian reflector.

44. A 35-cm reflector, f/5.6, mounted as a 'Dobsonian'. This is a very simple, easy to make mounting, but cannot be easily driven, and is suited to deep-sky rather than planetary or lunar observation, as it is difficult to use a high magnification. This telescope was used by Nigel Bannister.

x 192, though every observer will have his own particular tastes, and much will depend upon exactly what he wants to study.

The only function of a tube is to hold the optics in the right positions. Refractors always have solid tubes, but with a reflector the tube may be a skeleton, or perhaps partly enclosed. Also, tubes may be made of wood, metal or even plastic.

Because the main telescope will have a relatively small field of view, it is useful to have a *finder* attached to the side of the tube. This will be a small refractor, with low magnification but a wide field of view. When you set out to observe any particular object, bring your target into the finder-field simply by aiming it; when the object has been brought into the centre of the finder-field, it will be visible in the main telescope, always provided that everything is properly lined up. This is convenient even for a refractor, and for a Newtonian reflector it is almost a necessity. Of course, a permanently-mounted, driven equatorial telescope will have setting circles, which can be used to direct the telescope to any desired RA and declination, but setting circles cannot be used with altazimuth mountings apart from those of the ultra-modern computerised variety.

This may be the right place to make a few comments about the choice of one's first telescope. Generally speaking, the minimum useful aperture is 3 inches (7.6 cm) for a refractor, and 6 inches (15.2 cm) for a Newtonian reflector. Smaller telescopes can be bought in many camera-shops, and are supplied by many dealers, and these small instruments – such as 2-inch (5-cm) refractors and 4-inch (10-cm) Newtonians – look neat and attractive; but, frankly, they are a

which mirrors are used together with special correcting plates. They are compact, powerful and highly effective, but they are also expensive.

Much of what we have already said about refractors applies equally to reflectors. If, for example, we have a 6-inch (15-cm) reflector with a focal length of 48 inches (122 cm) the focal ratio will be f/8; we might then choose eyepieces with focal lengths of 1 inch (2.54 cm), giving a magnification of 48; a ½-inch (1.3 cm), giving x 96; and ¼-inch (0.6 cm), giving

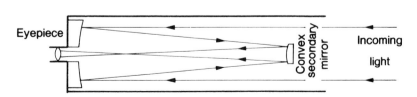

45. Principle of the Cassegrain reflector.

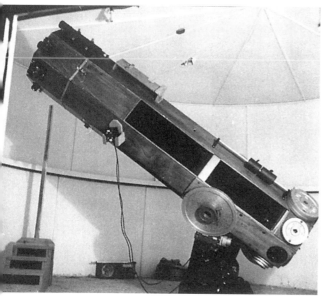

46. The 39-cm Newtonian reflector at Patrick Moore's observatory in Selsey. The octagonal wooden tube is only partially enclosed (to avoid tube currents) and the head is rotatable. There is an ordinary electrical drive, and several finders. The observatory has a rotatable roof with a slit.

Refractors are easy to use and need little maintenance; up to 3-inches aperture they are very portable, and anyone who is interested in the Sun will certainly prefer them. Reflectors are more temperamental, and the mirrors need regular attention, but against this a reflector will be usable with a higher magnification than a refractor of the same price. As always, it all depends upon which branches of astronomy interest you most.

Grinding a lens is a task for the expert. It used to be possible to make small refractors at negligible cost, by buying old spectacle-lenses and fitting them in cardboard tubes, but this is not so easy as it used to be, because the lenses are less common – though it may be worth inquiring at your local optician's shop.

On the other hand, it is quite possible to make an adequate reflector, either by buying the optics and mounting them, or else constructing the whole instrument – optics and all. It is laborious and time-consuming, but it can be done, and of course it cuts down the cost. Books such as *Small Astronomical Observatories* (Springer) will show you how to do it – but be prepared for many failures before you manage to make a really good telescope.

waste of money. They have poor light-grasp and small fields of view, and most of them are unsteady as well. In particular, beware of the telescope which is advertised as 'magnifying 250 times', or something of the sort. It is aperture which matters; and if the telescope is claimed to bear more than a maximum of x 50 per inch of aperture, then avoid it completely.

Second-hand telescopes can be found, but always take care before making a purchase – and if at all possible, take the advice of an expert. If an adequate telescope is to be bought new, go to a reputable supplier; and even then, be on your guard. You must be prepared for an initial outlay of £400 or so, but at least the cost is non-recurring, and it does not seem so very much when compared with, for example, the cost of a couple of rail tickets between London and Edinburgh!

47. Run-off observatory for Patrick Moore's altazimuth 32-cm reflector at Selsey. The shed is made in two parts, each of which runs back on rails.

Small telescopes can be moved around. Larger instruments are better set up permanently, probably in observatories, which may be of various designs (the run-off shed is very convenient, and easy to build). Naturally there are problems, and if there are any high trees around it is always found that they lie in the most inconvenient directions possible, while street lights produce other hazards. Generally, all that can be done is to accept the situation and make the best of it.

If you do not want to spend £400, or to make a reflector, the obvious alternative is to invest in binoculars, which are very useful astronomically and will show you a great deal. A pair of binoculars is made up of two small refractors, joined up and working together. They are classified according to their magnification and the diameters of their objectives: thus a 7 x 50 pair will have a power of 7, with each objective 50 mm across. A satisfactory pair of binoculars can be bought for £50, perhaps less. We would not recommend a magnification of above x 12, because the field of view will be small and the binoculars will be too heavy to hold steady without some form of mounting – as well as being much more expensive. But in any case, binoculars are very much better than very small telescopes.

Finally, remember that today there are many astronomical societies. Whether or not you are going in for GCSE, we suggest that you should join one. You will make many new friends as well as giving yourself a lifelong hobby – and you will certainly find somebody who is willing to let you look through his telescope.

Questions

1. Using diagrams, explain the principle of
    (a) An astronomical refractor,
    (b) A Newtonian reflector.
       What are the advantages and disadvantages of each type?
2. Give the magnifications produced in the following cases:
    (a) A 6-inch (15-cm) f/8 reflector with a ½-inch (1.27-cm) eyepiece.
    (b) A 3-inch (7.5-cm) f/10 refractor with a ½-inch (1.27-cm) eyepiece.
    (c) An 8-inch (20-cm) f/8 reflector with a ¼-inch (0.6-cm) eyepiece.
    (d) A 4-inch (10-cm) f/12 refractor with a 1-inch (2.5-cm) eyepiece.
    (e) A 3-inch (7.5-cm) f/12 refractor with an ⅛-inch (0.3-cm) eyepiece.
       One of these cases is not to be recommended. Which? And why not?
3. You have a 6-inch (15-cm) reflector with a focal length of 48 inches (122 cm). If you were choosing three eyepieces to use with this telescope, what would be their focal lengths? Give your reasons.
4. Two telescopes, X and Y, are both optically good. X is a refractor of 3-inches (2.5-cm) aperture, with a focal length of 48 inches (122 cm); it has an altazimuth tripod mounting. Y is a reflector of the Newtonian type, aperture 6 inches (15 cm), and has an equatorial mount on a tripod. Each telescope is provided with an eyepiece of focal length ½ inch (1.27 cm).
    (a) What magnification will be obtained with each telescope?
    (b) Which telescope will show the fainter stars – and why?
    (c) If you had a choice between these telescopes for your own use, which would you choose, and why?

5. (a) If you had a choice between 7 x 50 binoculars and an altazimuth 2-inch (5 cm) refractor, which would you prefer, and why?

(b) What are the disadvantages of binoculars with a magnification of greater than about x 12?

6. Calculate the percentage reduction of the light received by a Newtonian reflector with a 10-cm mirror, due to a 2-cm obstruction in the tube caused by the flat.

Practical work: making a small refractor

This is no place to go into details of making a reflector, but making a small refractor out of cardboard tubes and simple lenses is an official GCSE project, so that we must say something about it (Fig. 48).

The only materials needed are two lenses, some cardboard tubes, glue, brown paper, and pieces of wood. None of these will present any problems apart from the lenses. These are not nearly so easy to obtain as they used to be, but your local optician may help. The saving grace is that if you do find any, they will not cost much. £5 should cover the cost easily.

We need two lenses; one to serve as the object-glass, about 2 inches (5 cm) in diameter and about 30 inches (76 cm) focal length, and the other to act as the eyepiece, less than 2 cm in diameter and with a focal length of between 2½ and 5 cm. (A jeweller's eyepiece will do quite well.)

Next, we need two tubes. The longer of the two should have the same outside diameter as the object-glass, so that the OG can fit neatly on to the end. If the focal length of the OG is 30 inches, the tube should be about 28 inches long. Tube No. 2 carries the eyepiece; it must slide inside tube No. 1 (see Fig. 48), and to allow for focusing it should slide easily, though not freely enough to fall out. This tube should be about 16 inches (40 cm) long.

Very probably your local bookshop or newsagent will be able to provide tubes; if not, they can be made by using layers of brown paper glued together round a thick broomhandle.

Cut about 2 inches (5 cm) squarely off the main tube, and save the ring. Next cut a strip of brown paper 4 inches (10 cm) wide. Lay it down and paste it with glue. Then take tube 1, and, laying 2 inches of one end on to the end of the glued paper, roll the tube along so that the paper wraps round the end and sticks. Continue rolling until the cap around the end of the tube is reasonable thick. Leave it to dry.

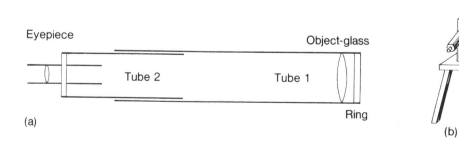

48. Making a simple refractor: (a) tube; (b) mounting.

Taking the ring that has been cut off tube 1, put it on to a piece of card and run a pencil round the outside. Cut out the disk you have drawn, put a 1½-inch (3 to 4-cm) disk on its centre, draw round the disk, and remove the centre circle of cardboard; we now have a ring which is to be used as a stop. Drop the stop into the cap on the end of tube 1. Next, put the lens into the cap; to fix it, put a little glue on the surface of the ring which has been cut off tube 1, and push it into the cap so that it rests against the lens.

To fit the eyepiece, simply jam it into the end of tube 2. If it does not fit, you will have to make adjustments, but this is easy enough. Make sure that everything is square-on. If either the eyepiece or the OG is lying at an angle, the telescope will be useless.

The effective aperture will be only about 1½ inches, because we have left only the centre of the lens clear of obstruction. If the focal length of the eyepiece is 1¼ inches, and the focal length of the object-glass is 30 inches, the magnification will be 30 divided by 1½ = x 20.

The mounting is a sheer problem in carpentry, and the diagrams here speak for themselves; but it is absolutely essential to have a mounting, and to make it sturdy, particularly as the field of view will be small.

The telescope will be far from perfect. Because the OG is a single lens, it will give a good deal of false colour, and it will certainly be less effective than binoculars! But it will be better than nothing at all, and you will at least have the satisfaction of knowing that you have made it yourself.

For an alternative construction project, see the Sundial section in Chapter 6.

9
Non-Optical Observing

Already in this book we have mentioned many times the results and discoveries that have resulted from observing in regions of the spectrum other than visible light, which after all makes up only a very limited part of the spectrum (as you can see from the diagram in Chapter 7). The new insights obtained have been simply stunning, and there is much more to come, but of course we can give only the briefest of glimpses below.

The first of the new fields to get started was radio astronomy, which may be the only science to have started by accident! In 1931, radio engineer Karl Jansky was investigating a mysterious source of 'static' and discovered that the interfering radio waves were in fact coming from the sky. Helped by advances in radar technology during the war, by the late 1940s results were beginning to flow thick and fast. The discovery of radio waves from the Sun in 1942 was quickly followed up and the great sunspot of 1946 (one of the largest ever observed) was discovered to be a powerful source. In the same year, Cygnus A, the first source beyond the Solar System was found and it was followed by many more.

It is worth saying something at this point about the way a radio telescope works. The principle is much the same as that of an ordinary reflecting telescope; radio waves (but remember radio waves and light are just two types of electromagnetic radiation) are collected by a parabolic dish and brought to a focus above the dish. Unlike a Newtonian reflector, there is no need for a secondary mirror and the signal is detected at this focal point by an electronic sensor (the principle behind this is no different to that on your radio set at home!). The major difference, as is immediately obvious from looking at the photographs of radio telescopes, is one of scale. Remember light has a *wavelength* of between 400 and 700 nanometres, while radio can have a wavelength from a few tens of centimetres to many tens of metres! This means that in order to collect a similar number of waves and so achieve a similar resolution, a radio telescope must be much larger. In fact, the resolution of single dish radio telescopes has lagged far behind their optical counterparts; it is simply not possible to construct large enough dishes and so a new method was sought.

What was found was the principle of interferometry; where the signal from two or more telescopes is combined to give a much greater angular resolution. Let us take, for example, MERLIN (the Multi-Element Radio Link Interferometer Network) in the UK. Here, the telescopes at Jodrell Bank (near Manchester) are linked with several other telescopes as far away as Cambridge to provide a telescope with an effective size of several kilometres. Other examples include the VLA (Very Large Array), which combines some eighty telescopes in the desert of New Mexico. Modern communications have made further advances possible, and it is now quite common for the large Lovell telescope at Jodrell to link up with telescopes in North America, providing resolution below 0.001 arc second (far better than any optical system to date, and equivalent to resolving a cricket ball 16,000 km away!).

Radio has provided us with discoveries that simply couldn't be obtained in any other way. While ordinary visible light may be blocked by dust lying in the way, radio waves are able to penetrate through, and so we owe all of our knowledge of regions such as the centre of our galaxy to radio. Other objects are especially active in the radio region; such as pulsars (the name derives from 'PULSAting Radio Source') or quasars which are dealt with elsewhere. Another technique, which provided us with our first really accurate idea of solar system distances, is to use the radio telescope as effectively a giant RADAR system, and in this way signals were bounced off, for example, the surface of Venus and the time taken for echoes to return recorded. This information, taken with knowledge of the speed of light, gives an excellent idea of the distance, in the same way that knowing the approximate speed of sound and counting for the time between lightning and thunder gives you an idea of the distance of the storm.

Before we move on from radio astronomy, we must mention one problem which may threaten the future of the science. Space in the radio bands is an extremely precious commodity with the advent of modern communications such as mobile phones, and efforts are being undertaken (by the International Astronomical Union, among others) to ensure that various important regions, such as the 21 cm wavelength where hydrogen emission is found, are kept clear. It may well be a losing battle, however (even emission from microwave ovens has been found straying into 'protected' bands) and the long term solution may well be to place a telescope on the Moon. Of course, there would then be the opportunity for an even longer baseline and still better resolution, but we must move on.

The great advantage of radio observing over the other regions of the spectrum is that it can be easily done from the Earth's surface. Infra-red radiation, however, is absorbed by water vapour in the Earth's atmosphere, which causes many difficulties. This field of study goes back to the days of William Herschel, who in 1801 discovered IR radiation from the Sun by placing a thermometer just beyond the red end of a solar spectrum, but little more was done until the 1960s when the first surveys of the sky were undertaken. By 1979, UKIRT (see Chapter 7) was up and running, by which time planning for the first space-based observatory was under way. The Infra-Red Astronomical Satellite (IRAS) was launched in January 1983 and functioned perfectly until November that year, when it ran out of coolant. It surveyed 97 per cent of the sky and catalogued almost a quarter of a million sources: from comets (one of which, IRAS-Alaki-Alcock, it jointly discovered) to dust clouds around stars such as Vega and Beta Pictoris. This work is still being followed up by newer instruments such as the Hubble Space Telescope, capable of observing in some of the IR regions of the spectrum, and the European Space Agency's Infrared Space Observatory. Many interesting objects, in particular star forming regions or the

49. IUE: the very successful International Ultra-violet Explorer satellite.

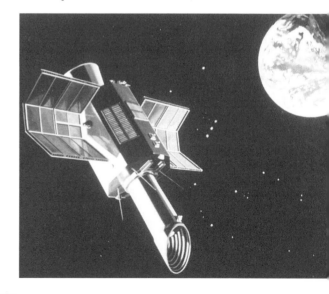

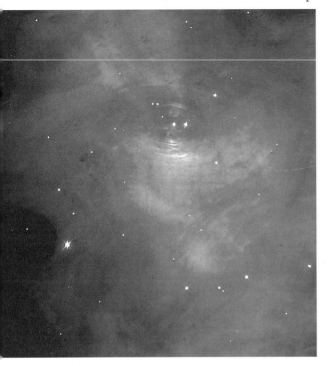
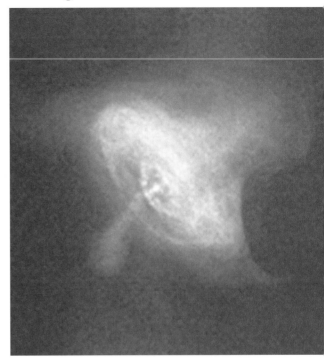

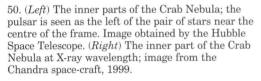

50. (*Left*) The inner parts of the Crab Nebula; the pulsar is seen as the left of the pair of stars near the centre of the frame. Image obtained by the Hubble Space Telescope. (*Right*) The inner part of the Crab Nebula at X-ray wavelength; image from the Chandra space-craft, 1999.

galactic centre where the dust obscures visible light, are best observed in IR.

The remaining regions of the spectrum have not been neglected either. Ultraviolet astronomy is impossible from the surface of the Earth, and so all the research has to be done from space. The most successful mission to date has been the International Ultra-violet Explorer (IUE), which, with a planned life expectancy of three years after launch in 1978 was still functioning well and returning useful results well into the nineties. In X-rays, observations have been led by the German telescope ROSAT, but at the time of writing results are just beginning to come in from NASA's

Chandra observatory. X-ray observations specialise in providing excellent views of some of the universe's more exotic and energetic objects, such as systems like the Crab Nebula, the remains of a star which went supernova just over a thousand years ago. Finally, gamma ray telescopes such as the Compton Gamma Ray Observatory (another NASA space based telescope) have detected radiation from pulsars, from very active galaxies and also provided us with one of our biggest mysteries; gamma ray bursts. These extremely powerful sudden bursts are at present the source of much debate as to their origin, and only time will tell.

As we said before, it is not possible to give much more than an overview of what is being done, but it should at least be clear that only by studying in all the available regions of the spectrum can we hope to fully understand everything we observe.

10
Gravitation

Gravitation may be said to be the most important force in the whole universe. It is the force of mutual attraction – every body attracts every other body. It is gravitation which keeps us on the surface of the Earth, and which keeps the Earth and the other planets in orbit round the Sun; it applies equally to the distant star-systems thousands of millions of light-years away from us.

Though it was Newton who laid down the laws of gravitation, the story must begin with Kepler, who first worked out the way in which the planets move round the Sun. His famous Laws of Planetary Motion, published between 1609 and 1618, are as follows:

1. The planets move in elliptical orbits. The Sun lies at one focus of the ellipse, while the other focus is empty.
2. The *radius vector*, or imaginary line joining the centre of the planet to the centre of the Sun, sweeps out equal areas in equal times.
3. The squares of the sidereal periods of the planets are proportional to the cubes of their mean distances from the Sun.*

Law 1 is straightforward enough. Fig. 51 shows an ellipse; the two foci are marked F and F', and the maximum diameter of the ellipse, AB, is termed the *major axis*. The ratio between the distance between the foci (F-F') and the major axis (AB) is a measure of the *eccentricity* of the ellipse. If the two foci are in the same place, the eccentricity is zero – in other

* The mathematical formula is $(T_1/T_2)^2 = (r_1/r_2)^3$.

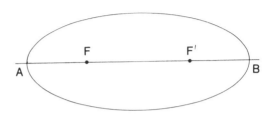

51. Ellipse: AB = major axis; F and F' = foci.

words, the ellipse has become a circle. If the eccentricity is 1, we have an open curve or *parabola*; and if it is greater than 1, we have a *hyperbola* (Fig. 52). The eccentricity of the Earth's orbit is 0.017, so that it is not very different from a circle. This is also true of the other planets (even with Pluto, whose path is the least circular, the eccentricity is no more than 0.248), but comets often have very eccentric orbits, and some of them move in parabolic or even hyperbolic paths, so that when they have passed round the Sun they never come back at all.

Law 2 is also easy to understand. In Fig. 53 – the left-hand diagram suitable for a planet, the right-hand diagram for a

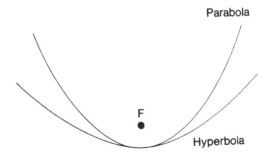

52. Parabola and hyperbola.

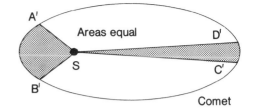

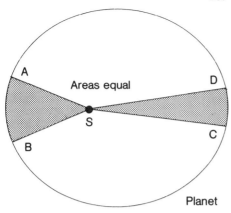

53. Kepler's Second Law. *Left*: area ASB must be equal to area CSD; *right*: area A'SB' must be equal to area C'SD'.

comet – the body moves from A to B in the same time that it takes to move C to D. This means that with S representing the Sun, the area ASB must be equal to the area CSD, and A'SB' must be equal to C'SD'. In fact, the body moves fastest when it is closest to the Sun, at perihelion, and slowest when it is furthest out, at aphelion. It follows that planets which are close to the Sun move more quickly than those which are further out. Mercury (mean distance from the Sun, 58,000,000 km) has an average orbital velocity of 48 km per second; Venus, at 108,000,000 km, 35 km per second; the Earth, at 150,000,000 km, 29.8 km per second, and so on out to Pluto, whose mean distance from the Sun is 5,900,000,000 km and which crawls along at an average rate of only 4.7 km per second.

Law 3 is rather more complicated. Essentially, it means that there is a definite relationship between a planet's sidereal period – that is to say its 'year', the time taken for it to go once round the Sun – and its mean distance from the Sun. This has proved to be very useful in measuring the length of the astronomical unit, or Earth-Sun distance. We can measure the distance of the planet Venus, for example (the modern method is to bounce radar pulses off it, and see how long they take to go there and come back)

and of course we know the sidereal periods of both Venus and the Earth, so that it is a fairly simple calculation to work out the only remaining unknown in the equation – i.e. the distance between the Sun and the Earth.

Note that Kepler's Laws apply equally to satellites, natural or artificial, moving round the Earth and other planets – and also to binary star systems. Gravitation is universal.

Now let us turn to Newton, whose great book, the *Principia*, appeared in 1687. Newton's laws state that:

1. Every body continues in its state of rest, or uniform motion in a straight line, unless it is affected by some external force.
2. The rate of change of momentum is proportional to the applied force, and takes place in the direction in which the force acts.
3. To every action there is an equal and opposite reaction.

Most people have heard the story of Newton and the apple, which, unlike most other tales of its kind (such as Canute and the waves) is probably true. When at his Lincolnshire home during the Plague period, Newton saw an apple fall from a tree, and this simple event started a chain of thought in his mind. He realised that the force acting on the apple is the same as the force which keeps the Moon in its orbit. Why, then, does the Moon not fall down? The answer is: Because it is moving.

If you whirl a conker round on the end of a string, it will keep on moving, because it is being pulled by the string. If the string breaks, the conker will fly off at a tangent (Fig. 54). If we imagine that the conker represents the Moon, we can see that it also would move off at a tangent if it were not being pulled on by the Earth.

In the next diagram (Fig. 55) the Moon is shown in its orbit. If it were not being pulled upon, it would travel from M in the direction M1. Actually, it is 'falling', and instead of reaching M1 it arrives at M2. It goes on 'falling' round the Earth all the time, but it never comes any closer to us. In one minute it 'falls' by approximately 4 m. The diagram given here is hopelessly out of scale, but it is good enough to show the general principle.

Rather surprisingly, it is not correct to say baldly that the Moon moves round the Earth. Newton established that a body

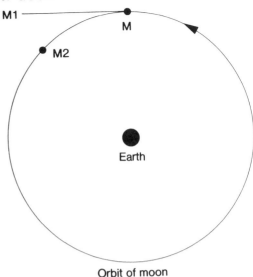

55. Movement of the Moon. If the Moon were not being pulled upon by the Earth, it would move from M to M1. Because of the Earth's gravitation, it 'falls' and reaches M2 instead.

will behave, gravitationally, as though all its mass were concentrated at a point in the centre of the body, and the Moon's mass is quite considerable; it is 1/81 of the mass of the Earth, so that if the Earth were put in one pan of a gigantic pair of scales you would need 81 Moons to balance it. Strictly speaking, the two bodies revolve round their common centre of gravity, or *barycentre*. This can best be explained by picturing a pair of gymnasium dumb-bells (see Fig. 56). If the two bells are equal in weight, you can balance them midway along the joining bar (a). If one bell is heavier than the other, the balancing point, or centre of gravity of the system, will be closer to the heavier bell (b). With the Earth-Moon system, the balancing point is actually inside the Earth's globe (c), and this is the point around which both bodies move. For most purposes it is good enough to say that the Moon revolves round the Earth in a period of 27.3 days; but the difference is mathematically important.

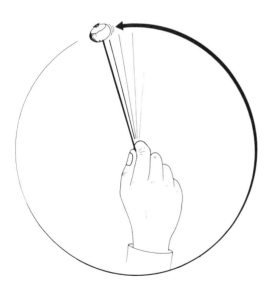

54. Whirling a conker. If the string breaks, the conker will fly off at a tangent.

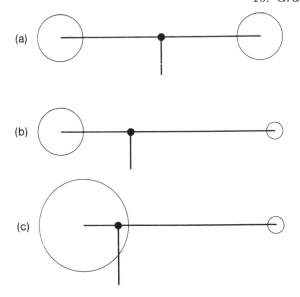

56. Balancing point for the bells of a dumb-bell system. (a) Equal bells; balancing point midway. (b) Unequal bells; balancing point nearer the heavier bell. (c) Very unequal bells; balancing point inside the more massive bell – as with the Earth-Moon system.

Note that it is mass which matters, not size. In these diagrams the heavier bell has been drawn the larger of the two, for the sake of clarity, but the balancing point or centre of gravity will always be displaced toward the more massive bell even if it happens to be the smaller of the two – as in some double star systems.

There are other complications too. For instance, the planets pull upon both the Earth and the Moon, and cause disturbances or *perturbations* which are large enough to be measurable. But of course the Sun, with its tremendous mass, is the controlling body; in fact the Moon may be said to have a planetary-type path, and its orbit is always concave to the Sun (Fig. 57).

The more massive a body, the greater its gravitational pull; but the surface gravity of a planet (or a satellite) depends not only upon its mass, but also upon its size. Remember, gravity acts as though it were concentrated at a point in the centre of the globe, and the further away you are

from the centre the weaker will be the pull. Go to the Moon, and you will have only one-sixth of your Earth weight, because although the Moon is much smaller than the Earth it is also much less massive. But consider the two planets Mars and Mercury. Of the two, Mars is the more massive; but it is larger than Mercury, so that a man standing on the surface of Mars is further away from the centre of the planet than a man standing on the surface of Mercury. The result is that the surface gravities of Mars and Mercury are the same (0.38 that of the Earth).

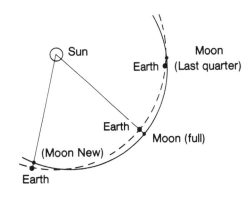

57. The Moon's orbit; it is always concave to the Sun.

The greater your distance from the centre of the Earth, the weaker will become the Earth's pull. Because the globe is slightly flattened, the poles are slightly closer to the centre of the Earth than the equator (Fig. 58), which is why you 'weigh' slightly more if you are standing at the north pole than you do if you are standing on the equator – though we can assure you that the difference is not very marked!

And this leads us on to Newton's *inverse square law*, according to which two bodies will pull upon each other with a force which is directly proportional to the product of their masses, and inversely proportional to the square of the distance

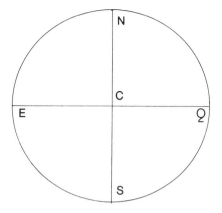

58. C is the centre of the Earth; EQ = the equator; NS = the prime meridian. Obviously EQ is longer than NS, so that if you stand on N or S you are nearer the Earth's centre and weigh more. Needless to say, this has been drawn wildly out of scale, for the sake of clarity.

between them.* Yet although the pull of gravity becomes weaker when the bodies move further away from each other, it does not vanish completely. In theory it will never disappear, though over a sufficiently large distance it becomes too slight to be measured.

To demonstrate the inverse square law, let us assume that we have two planets moving round the Sun, one at a distance of 2,000,000 km and the other at 5,000,000 km. (Of course, no planet is as close-in as this, but we are making things as simple as we can.) $2^2 = 4$; $5^2 = 25$. The force on the planets will be in the ratio of 1/4 to 1/25, so that the force on the more remote planet will be only 4/25 of that on the nearer planet.

Finally, in this admittedly rather rambling account, we must say something about the tides, which are due to the

gravitational pulls of the Moon and, to a lesser extent, the Sun.

The Moon is only 384,000 km from us on average, and, as we have seen, it is a massive body even though it is by no means the equal of the Earth. The Moon's pull tends to heap up the water in our oceans, and to keep a bulge of water below it (Fig. 59), producing a high tide. There is also a high tide on the opposite side of the Earth, because the water in the oceans is further from the Moon than the land, and a heap is produced. As the Earth spins, the two tidal bulges do not spin with it; they tend to stay in the same positions, so that in theory every place ought to have two high tides and two low tides each 24 hours.

Of course, this diagram is hopelessly over-simplified. For one thing, the Earth is not surrounded by a uniform shell of water; also there is a time-lag, so that the heap is not directly under the Moon, and the whole phenomenon of the tides is very complicated indeed. But the main principle is clear – and note that there is also a 'land tide', though it is so slight that in everyday life it cannot be noticed.

The Sun also has tidal effects on the oceans, but solar tides are weaker than lunar ones because the Sun is so much further away from us. When the Sun and Moon are pulling in the same direction, or in exactly opposite directions, we have strong or *spring* tides; when the Sun and Moon are pulling at right angles to each other, we have the weaker *neap* tides. As can be seen from Fig. 60, spring tides occur when the Moon is either new or full (they have nothing to do with the season of spring).

We have said nothing here about the true nature of gravitation, because it certainly does not come into any GCSE syllabus. But at least we hope that what we have said helps to explain how it acts.

*If F stands for the force, m_1 and m_2 are the masses of the bodies concerned, d is the distance between their centres, and G is a term known as the gravitational constant, then

$$F = \frac{G\, m_1\, m_2}{d^2}$$

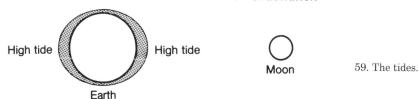

High tide   High tide

Earth

Moon

59. The tides.

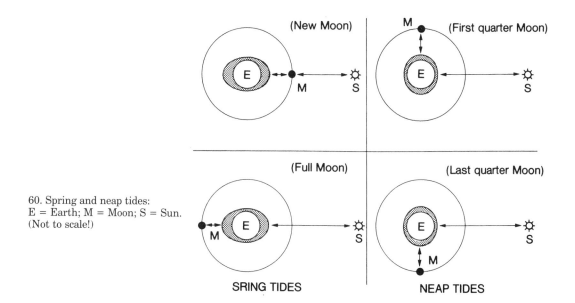

(New Moon)

(First quarter Moon)

E — M — S

M — E — S

60. Spring and neap tides:
E = Earth; M = Moon; S = Sun.
(Not to scale!)

(Full Moon)

(Last quarter Moon)

M — E — S

E — S

M

SRING TIDES

NEAP TIDES

Questions

1. State Kepler's Laws. Where necessary, illustrate your answer by means of diagrams.
2. Explain, using a diagram, why the Moon may be said to be 'falling' continuously toward the Earth. Why does it not fly off in a straight line?
3. Describe the manner in which the Earth and the Moon move round the barycentre. What difference would it make if the Earth and the Moon were equal in mass?
4. Give an explanation of the phenomenon of the tides, explaining the terms 'spring tide' and 'neap tide'.
5. Venus moves round the Sun at a mean distance of 108,000,000 km, the Earth at a mean distance of 150,000,000 km. What is the ratio of the force of gravity on the Earth as compared with that on Venus?
6. Given that the distance between the Earth and the Sun is 1 a.u. (astronomical unit), and the orbital period is 1 year, calculate the orbital period of Mars, whose distance from the Sun is 1.5 a.u.

11
Artificial Satellites

The idea of space-travel is not new. It goes back many centuries, but only in our own time have we learned how to achieve it. Many people can remember the time when the idea of sending a rocket to the Moon was treated as nothing more than a joke!

Probably the earliest science-fiction story still widely read is the Moon-voyage novel written by the French author Jules Verne in the 1860s. Of course, it is hopelessly old-fashioned now, but it is worth recalling here, because we can learn a great deal from it. In particular, Verne had the right idea about what we call *escape velocity*.

Throw an object upwards, and it will rise to a certain height and then fall down. Throw it at a greater speed, and it will rise higher before falling back. If you could throw it upward at a speed of 11.8 km per second, which works out at slightly more than 42,000 km per hour, it would never come back at all. The Earth's gravity would not be strong enough to hold it, and the object would escape into space – which is why 11.8 km per second is the Earth's escape velocity. (There is no point in trying it. The world's greatest fast bowlers cannot send down a delivery as fast as 160 km per hour!)

In Verne's novel, the projectile containing the three space-travellers was fired toward the Moon at full escape velocity by means of a huge cannon. In theory, this was sound enough, but there were two points which Verne either did not know or else chose to ignore. First, moving quickly through the air sets up heat by friction, and if a projectile were fired at full

escape velocity it would have been destroyed even before it left the barrel of the cannon. Secondly, the shock of a sudden departure would have caused quite a jolt, to put it mildly. The luckless travellers would have been turned into jelly even if they had managed to avoid being burned up.

Before going any further, there is another important point to be made about escape velocity. Air is made up of vast numbers of atoms and molecules, all flying about at high speeds. If a particle can travel fast enough, it will escape. Luckily for us, the main particles in our air cannot achieve escape velocity, but things are very different on the Moon, which is much less massive than the Earth and has a lower escape velocity – only 2.3 km per second. This is not enough. If the Moon ever had an atmosphere, it could not hold it down; the lunar 'air' leaked away into space, and by now there is none left. On the other hand the massive planet Jupiter, with an escape velocity of over 60 km per second, has been able to hold down all its original gases, including the very lightest of all – hydrogen. This is why Jupiter's atmosphere is now made up chiefly of hydrogen, and to us would be unbreathable.

Verne's space-gun being rejected, we must turn to the principle of the rocket, first described in detail by the Russian pioneer Konstantin Tsiolkovskii, toward the end of the nineteenth century. (Tsiolkovskii's name is pronounced 'Ziolkovsky', which would be a more sensible way of spelling it.) Once again we come back to Newton, whose *principle of reaction* stated that 'every action has an

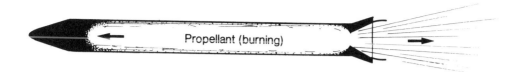

61. Principle of the rocket. The gases escaping from the exhaust propel the rocket in the opposite direction; there is no need for any surrounding air – the rocket is 'pushing against itself' according to Newton's principle of reaction.

equal and opposite reaction'. As Tsiolkovskii realised, space-travel is only possible in a vehicle which can operate without having air around it – and there is no air above a height of a few hundreds of kilometres.

Consider the Guy Fawkes rocket which we fire every November the Fifth. It is made up of a hollow tube filled with gunpowder. When you 'light the blue touch-paper and retire immediately', the gunpowder starts to burn. As it does so, it produces hot gas, which tries to escape from the tube. It can do so in only one direction – where the touch-paper has been burned away. Therefore it rushes out of the rocket in a concentrated stream, and in so doing it propels the tube in the opposite direction. It is, so to speak, 'kicking against itself' (Fig. 61.).

There are two simple experiments which will show what is meant. Stand on a roller-skate or a skateboard, and jump off; the skateboard will move in the opposite direction to your jump – and if you could carry out this experiment in a vacuum (assuming that you could breathe there!), it would still work. If you blow up a balloon and then release the neck, it will fly across the room. True, it twists and turns, but this is because it is pushing against the air; in vacuum, it would fly in a straight line (Fig. 62).

Now let us come back to the space-rocket. Here we have two liquids, a propellant and an oxidant, which are fed into a combustion chamber. When they

combine, they react together; hot gas is produced and sent out through the exhaust – so that the rocket flies. Note that it, too, is moving because of the principle of reaction. Atmosphere is not needed; in fact it is actually a nuisance, because it has to be pushed out of the way. Therefore, the rocket will work very well in space.

Also, a liquid-propellant of this kind can be controlled. It can start off slowly, and work up to full velocity only when it is so high up that it is no longer in danger of being burned away by friction against the atmosphere.

Tsiolkovskii himself was purely a theorist; he never fired a rocket in his life. The first liquid-propellant rocket was sent up by an American, Robert Goddard, in 1926; and though it was a very modest affair, it was the direct ancestor of the space-ships of today. It is, unfortunately, true that during the war rockets were developed as weapons of war, by the German team headed by Wernher von Braun, but it is also true that if there had

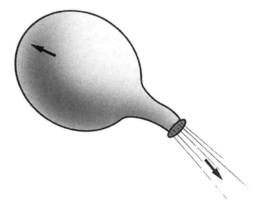

62. Balloon in flight – as the air is released, the balloon shoots across the room.

been no military connection the progress of rocketry would have been much slower than it actually was.

A modern rocket launcher is a complex affair. Even liquid propellants have their limitations, and the procedure is to use a 'step-vehicle', in which several rockets are mounted one on top of each other. At first the large lower rocket does all the work; when it has used up its fuel it breaks away and falls back to the ground, leaving the second rocket to continue the journey by using its own motors. In theory there can be any number of steps, though the practical problems mount up quickly.

Scientific rockets were used both by the Americans and the Russians in the years following the end of the war, in 1945, but they had their drawbacks; they could not stay up for long, and no rocket could be used more than once, which made the whole programme very expensive indeed. Then, on 4 October 1957, came a major advance. Rather to the surprise of many Western observers, the Soviet Union launched the first artificial satellite, Sputnik 1. It was only about the size of a football, and it carried little apart from a radio transmitter, but it ushered in the Space Age.

The basic principle is to put the satellite into the top stage of a step-rocket, send it above the atmosphere (or at least that part of the atmosphere which is dense enough to cause serious friction), and then put it into an orbit round the Earth. Once this has been done, the satellite will not come down. It will behave in the same way as a natural astronomical body, and it, too, will obey Kepler's Laws.

In Fig. 63, let us imagine that there is a high tower poking out above the top of the atmosphere, and that we can fire vehicles off its top in a direction parallel with the Earth's surface. Vehicle A, sent out at low velocity, will hit the ground at point A'. Vehicle B, given a higher velocity, will travel as far as B'. Vehicle C, launched at 'circular velocity' of approximately 8

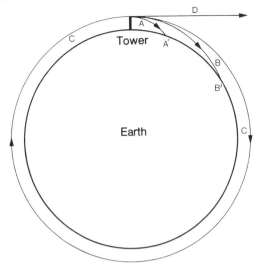

63. The tower protrudes from the atmosphere (needless to say, the diagram is out of scale). Vehicle A hits the ground at A'; B, at B'; C, at circular velocity, will not come down, and D, at escape velocity, will break free altogether.

kilometres per second, will never come down at all. It will 'fall', just as the Moon does; but, like the Moon, it will not return to the ground. Vehicle D, launched at escape velocity, will leave the Earth altogether.

Of course, this drawing is completely out of scale, and in any case the real launching procedure is to send the vehicle upward in a pre-computed curve instead of sending it vertically up and then altering its course by 90°. But the main idea is straightforward; the satellite will enter an elliptical orbit whose precise size and shape will depend upon its initial velocity and direction. (A circular orbit will seldom be found. After all, a circle is a special case; it is an ellipse with zero eccentricity.) Again following Kepler's Laws, the period taken to complete one journey round the Earth will depend upon the distance of the satellite from the Earth's centre; the greater the distance, the longer the period.

We know that the density of the

atmosphere falls off with increasing distance above the ground, and above about 250 km there is so little air left that the satellite is not noticeably affected by friction. However, if any part of the orbit lies within the denser atmosphere, as in Fig. 64, the orbit will shrink, because when near perigee (minimum distance from the Earth) friction will cause a braking effect. Finally, the satellite will drop into the denser part of the air, and will be burned away in the same way as a meteor.

A satellite is lit up by the Sun, and appears as a starlike point, but its movement against the background of real stars is noticeable; of course, the lower the orbit of the satellite, the quicker it seems to go. When a satellite passes into the shadow of the Earth it is eclipsed, and fades from view – a fact which has puzzled many unwary people, and has given rise to a whole crop of flying saucer stories!

Satellites have been put to many practical uses. First, they are used for communication; without them it would be impossible to have direct television links over great distances. When Test Matches are being played in Australia, you can see them on your television because satellites put into suitable orbits are used as relays. Radio and telephone links are of vital importance today, and many communications satellites have been launched. There are also many weather satellites, which can send back pictures of complete weather systems and have done a great deal to improve our forecasts. Other satellites study the Earth itself; for example they can detect areas of diseased forests, and have been of great use in helping geologists to identify regions where oil may be found. To list all their uses would take many pages; it is enough to say that they are tremendously valuable – which is in itself justification for spending money on them. (A great deal is heard about the cost of space research, but it is worth noting that in one year America spends much less upon

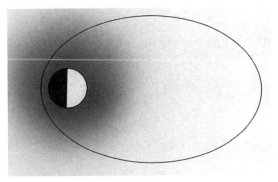

64. Satellite orbit. Here part of the orbit lies within the region where the Earth's atmosphere is dense enough to cause appreciable resistance. Therefore, the satellite will not be permanent. Its orbit will contract (i.e. the apogee distance will become less) until the satellite enters the denser air and is destroyed by friction.

space programmes than its people do upon chewing-gum.) For some years all the satellites were either Russian or American, but by now other nations have joined in. Space is truly international.

Now let us turn to purely scientific uses. One startling discovery was made in 1958 by the first successful United States satellite, the tiny Explorer 1, launched by a team headed by Wernher von Braun, who went to America after the defeat of Germany in 1945. Explorer's instruments found that there are zones of intense radiation encircling the Earth, now known as the Van Allen zones in honour of James Van Allen, who designed the equipment. Changed particles sent out by the Sun are trapped in the Earth's magnetic field, and are unable to escape. Major outbreaks on the Sun, known as solar flares, produce extra numbers of particles; the Van Allen zones become overloaded, so to speak, and the charged particles cascade downwards toward the magnetic poles, entering the upper air and causing the magnificent displays of coloured lights known as *aurorae* (Aurora Borealis in the northern hemisphere, Aurora Australis in the southern). From countries such as North Norway and Alaska, aurorae can be seen on many nights during the year, and they are fairly common from most of Scotland, though we see fewer aurorae in England because we are further away from the magnetic pole.

The Van Allen zones are of great importance in all studies of the Earth's *magnetosphere*, the extent of the terrestrial magnetic field, but without artificial satellites we would probably still be blissfully unaware that the trapped particles exist at all. Probably this was the most important single discovery made during the first years of space exploration.

We can also study cosmic rays, which are not really rays at all, but high-speed atomic particles coming in from deep space in all directions. They cannot reach the Earth's surface without being broken up by collisions with air-particles, so that again satellites are of vital importance.

Looking back at the electromagnetic spectrum, you will remember that most of the wavelengths are blocked by layers in the atmosphere, either totally or partially, so that to study them we need rockets, balloons or (preferably) satellites. Many satellites have been launched for specialised investigations. X-rays from space were first detected by a rocket as long ago as 1962, but there have been a number of X-ray satellites, such as Uhuru (1970), the so-called Einstein Observatory (which operated from 1978 to 1981), Exosat (1983 to 1986) and Chandra (launched 1999). X-rays are emitted by very hot bodies, and without satellites we would know very little about them. Ultra-violet radiations have been studied by satellites such as the Copernicus vehicle and the amazingly long-lived IUE or International Ultra-violet Explorer, which was sent up in 1978 and was still operating excellently in 1995. It was finally closed down on financial grounds. In 1983 came IRAS, the Infra-red Astronomical Satellite, which functioned for the best part of a year and provided a complete map of all the infra-red sources in the sky above a certain level of power; it

also detected cool material around certain stars which may indicate the presence of planet-forming material, or even planetary systems – a point about which we will have more to say when we come to discuss the stars. One of the most successful modern space ventures has been the Hubble Space Telescope, launched from Cape Canaveral in 1990; it has a 240-cm mirror, which is not large by modern standards, but since it operates from above the top of the main atmosphere the 'seeing' conditions are perfect all the time. The Hubble Telescope has been a great success; preliminary faults were put right by astronauts who carried out a servicing mission, and the results obtained have been outstanding. The telescope is still operating faultlessly, and is expected to do so for some years yet.

Not all these satellites are high enough to remain aloft indefinitely. Many spiral downward, by friction, and are destroyed,

65. The Hubble Space Telescope, seen from the Shuttle.

though others are safe; at an altitude of some 36,000 km a satellite moving above the equator will keep pace with the rotating Earth and will appear to keep in a fixed position in the sky (this is known as a geosynchronous orbit). But there are dangers too. There have been cases of satellites carrying nuclear equipment which have come down unexpectedly, scattering débris over a wide area, and it is also true that satellites can be used for military preparations – something which many scientists regret. But all in all, the advantages of artificial satellites far outweigh their disadvantages.

Manned satellites date from 1961, when the late Yuri Gagarin, of Soviet Russia, made the first journey round the Earth in a space capsule (Vostok 1). Gagarin's flight was a pioneering venture in every sense of the term, because neither he nor anybody else knew quite what to expect. For example, it was by no means certain that a space-craft would not be hit by solid particles (meteoroids) and damaged, and cosmic rays were another possible danger. But most important of all, Gagarin experienced weightlessness or *zero gravity*, which we cannot properly reproduce on Earth simply because there is no way in which gravity can be screened.

If you put a coin on a book, the coin will press down on the book; with reference to the book, it will be 'heavy'. Now drop the book. During the descent to the floor, the coin will cease to press upon the book; it will be falling, but the book will be falling away from underneath it, and there will be no pressure, so that with reference to the book the coin will have become weightless. (Exactly the same would be true if the two were travelling upward.) Much the same can be said of a space-traveller and his vehicle. If they are moving in the same direction at the same rate, the astronaut will not press down on his space-ship, and will seem to have no weight at all. This happens when the space-ship is moving round the Earth in free fall, in just the same way that the real Moon does. Note

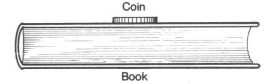

66. A coin on a book presses down on the book. If the two are dropped, there will be no pressure; with reference to the book, the coin will have become weightless.

that weightlessness is not the same as 'getting out of gravity', because both the astronaut and his space-ship are still well within the powerful grip of the Earth's gravitational field.

Zero gravity is a strange sensation; but according to those who have experienced it, it is not uncomfortable. (Gagarin himself once commented that he enjoyed it.) There are problems – for instance you cannot pour out drink, because a weightless liquid will not pour; you have to use a plastic bottle and squeeze it. Drop a piece of equipment, and it will not fall, but will simply stay hovering. There is no 'up' or 'down', and some astronauts have felt temporarily sick. But at least it is not harmful over short periods, even though an astronaut who has been in space for weeks or months takes some time to recover when he returns to Earth and becomes 'heavy' once more.

Whether there will be harmful long-term effects is not so certain; in particular, it seems that bones are affected and lose calcium. So far, astronauts (or, to use the Russian term, cosmonauts) have been in space for a continuous period of well over a year. Whether the human body will stand up to a prolonged flight to, say, Mars remains to be seen.

Gagarin's flight was the first of many. Among the major developments of the early 1960s were 'space rendezvous', when two vehicles were brought together and docked. If they are in the same orbit, they will not fly apart – any more than two ants crawling upon a bicycle-wheel will fly apart if the wheel starts to spin. For the same reason, an astronaut who puts on a space-suit and goes outside his vehicle will not drift away, though in most cases it is wise to use a safety-line.

Leaving travel to the Moon and unmanned probes to the planets for the moment, we must say something about space-stations such as America's Skylab, which was launched in 1973 and was manned by three successive crews. This was a purely scientific vehicle, and all sorts of experiments were carried out – including industrial research. (To give just one example, perfect crystals are best made under conditions of zero gravity.) Later came the Soviet station Mir, which was launched in 1986 and remained in orbit for over 14 years: it was manned by successive crews. The first sections of the International Space Station were launched in 1998-9. But if you are going to have a major space-station, you must be able to go up and down to and from it, and to use a new rocket every time is expensive and clumsy. This is why the Americans developed their Shuttles, which are recoverable vehicles. It has been said that a Shuttle takes off like a rocket, flies like an aircraft and lands like a glider.

The Shuttles took much longer to develop than had been expected. Then, after over thirty successful missions, came disaster; in 1986 the Shuttle *Challenger* exploded almost immediately after take-off, killing all nine members of its crew. The astronauts were not the first to die in space (four Russians had done so) but the tragedy caused a major delay in the whole United States programme, and not until 1988 were Shuttle flights resumed.

No doubt there will be many developments within the next ten years. There are encouraging signs of greater international co-operation in space; more stations are in an advanced stage of planning, and there is every hope that by the end of the century we will have a base on the surface of the Moon. Time will tell. But we would be prepared to bet that some readers of this textbook, now studying for GCSE, will themselves go into space. We have come a long way since the Russians launched their football-sized Sputnik over forty years ago.

Questions

1. (a) What is meant by escape velocity? What is the escape velocity of the Earth?
   (b) What effect does the escape velocity of a planet have upon the extent and composition of the planet's atmosphere?
   (c) If the Earth's escape velocity were ten times as great as it actually is, would we be able to live here? Give reasons for your answer.
2. (a) Using diagrams, explain the principle of the rocket.
   (b) Why can a rocket be used to send vehicles beyond the atmosphere, whereas a projectile fired from a space-gun cannot?
   (c) Explain the step-principle of launching probes, and show why it has to be used.
3. (a) Two satellites, X and Y, are in orbit round the Earth. X has an eccentric orbit, so that its height above sea-level ranges between 120 and 550 km; Y has an almost circular orbit at a mean height of 400 km above sea-level. Which satellite will remain in orbit for longer, and why?
   (b) Why is zero gravity experienced by an astronaut travelling round the Earth in free fall?
   (c) If an astronaut goes outside his space-craft while orbiting the Earth, will he fly away from it? If not, why not?
4. (a) Give four practical uses of artificial satellites.

    (b) What were the wavelengths studied by IRAS, IUE, and Chandra?

    (c) Why does a satellite in geosynchronous orbit have to be at a height of 36,000 km above the earth?

5. (a) What are the Van Allen zones?

    (b) What are aurorae, and how are they produced?

    (c) If you spent a year on a space-station without returning to Earth, would you feel any discomfort when you finally came down? If so, why?

12
Spectroscopy

Much of what we know about the universe derives from one technique alone; that of spectroscopy. It has made it possible for us to obtain much information, such as the chemical composition of the stars, that was thought for a long time to be forever beyond our reach.

We begin with Isaac Newton, famous primarily for his theory of gravity. He achieved much more besides, however, and one of his discoveries was that sunlight passed through a prism could be split into a rainbow showing all the colours of the spectrum. It wasn't until the spectrum was more closely examined in the early years of the nineteenth century that hundreds of dark lines were observed in this solar spectrum, firstly by Wollaston, and then later and in more detail by Fraunhofer (we refer to these lines today as the 'Fraunhofer lines'). The correct explanation was provided in 1859 by two chemists, Bunsen (of burner fame) and Kirchoff.

To understand the process that produces the lines, remember first that an atom can be thought of as having a number of negatively charged electrons orbiting a positively charged nucleus. These electrons can orbit only in certain 'energy levels' around the atom, and not in between them. The situation is similar to a ball on a staircase, where the ball can be on any one of the stairs (energy levels) but not in between them. Now, when light hits the atom, it gives energy to an electron which absorbs it and so the electron can jump to a higher energy level. This configuration is not stable, however, and so the electron falls back down, re-emitting the light. The crucial point is this; the difference in energy between the two levels determines the wavelength (colour) of the light, and different types of atoms have different values for these differences. Therefore, by observing the positioning of the lines in the spectrum, we can tell which elements are present.

In fact, the situation is a little more complicated. A gas at low pressure produces only a continuous spectrum; no lines at all. This is the case in the photosphere of the Sun, but above that lies the chromosphere, containing gas at higher pressure and so the effect discussed above comes into play. Normally the lines would be bright, but silhouetted against the bright background they appear dark. Other objects provide us with bright lines, but the basic principle is always the same.

A spectrograph is a reasonably straightforward instrument. Incoming light is passed through a prism (or a diffraction grating, which does the same job) just as in Newton's experiment. It is then

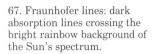

67. Fraunhofer lines: dark absorption lines crossing the bright rainbow background of the Sun's spectrum.

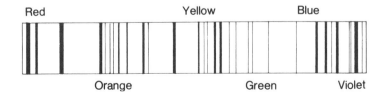

Red · Yellow · Blue

Orange · Green · Violet

possible to move along the spectrum with a detector, which may be simply a small telescope combined, of course, with the eye, or a more sophisticated electronic device. The wavelength of each line can be determined by comparison with a standard source, such as a light or laser of known wavelength. An observer will see many lines, but with some knowledge these may be organised into a series. For example, one particular set of lines forms what is known as the Balmer series and corresponds to emission from hydrogen.

One of the early successes for the technique was, as we have mentioned, determining the elements in the Sun through comparing the Fraunhofer lines with laboratory results. Once this had been done, one set of lines remained which Sir Norman Lockyer identified as a previously unnamed element. This was helium (named from the Greek word for Sun: helios) which was not to be discovered on Earth for another thirty years. It was not long before astronomers turned their new found technique toward more distant objects, and in particular the stars (see Chapter 19). Since then the spectrograph has been instrumental in every field of astrophysics; examples include discovering many double stars and, of course, the redshift of distant galaxies.

13
The Solar System

The Sun's family, or Solar System, may not be important in the universe as a whole, but it is certainly important to us, because it is our home in space. It is made up of the following bodies:

One star: the Sun

Nine planets: Mercury, Venus, the Earth, Mars, Jupiter, Saturn, Uranus, Neptune and Pluto.

The planetary satellites: Saturn has 18 known satellites, Jupiter 17, Uranus 20, Neptune 8, Mars 2, and the Earth and Pluto one each, with Mercury and Venus unattended.

Many thousands of asteroids or minor planets, most of which keep to the belt between the orbits of Mars and Jupiter.

Comets, which are true members of the Solar System, but most of which have very eccentric orbits.

Meteoroids. Shooting-star meteors are cometary débris; meteorites are more nearly related to asteroids – in fact there is every reason to believe that meteorites come from the asteroid belt.

A large quantity of thinly-spread inter-planetary matter, concentrated largely near the main plane of the Solar System. It is this matter, when illuminated by the Sun, which causes the glow we call the Zodiacal Light.

Data for the planets are given in the table on p. 80

Several things are apparent from this table, together with the plan of the Solar System (Fig. 68). First, the System is divided into two definite parts. There are four relatively small planets (Mercury to Mars), followed by a wide gap in which move the main asteroids, and then the giants (Jupiter to Neptune). Pluto seems to be in a class of its own; it may well be too small and lightweight to be ranked as a true planet, and its orbit is so eccentric that near perihelion it comes closer-in than Neptune, though the orbital inclination of 17° means that there is no fear of a collision. Secondly, apart from Pluto, all the planets move in orbits that are not very different from circles. And thirdly, again apart from Pluto, the orbits are very much in the same plane, so that if you draw a plan of the Solar System on a flat table you are not far wrong. This, of course, is why the planets seem to keep to the region of the Zodiac.

It used to be thought that the planets were pulled out of the Sun by the action of a passing star, but this attractive idea has so many mathematical weaknesses that it has had to be given up. It now seems certain that the planets were formed out of a 'solar nebula' – that is to say, a cloud of dust and gas associated with the early Sun. The planets which formed close-in (that is to say, out as far as Mars) lost much of their original light material, but the planets produced further out could retain it, which is why they contain so much hydrogen and helium. We know the age of the Earth to be of the order of 4.6 thousand million years, and this is also true of the Moon, so that there is no reason to doubt that this is also the approximate age of the entire planetary system.

The satellite families of the planets are very different from each other. The Earth,

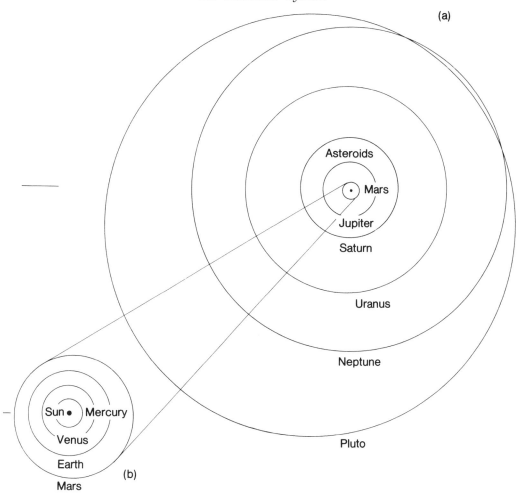

68. Plan of the Solar System. (a) Orbits of the outer planets, with Mars' orbit reduced to the same scale; (b) orbits of the inner planets.

of course, has only one natural satellite – the Moon – which is much the closest body in the sky. We are not certain about its origin. The old idea that it used to be part of the Earth, and broke away, has been abandoned by most astronomers; it may have formed in the same way as the Earth, at the same time and in the same region of space, so that the two bodies have always been associated. However, many astronomers now support the theory that in its early days the Earth was struck by a massive impactor, and the ejected material formed the Moon.

There can be nobody who is not familiar with the Moon's phases, or apparent monthly changes of shape. The diagram in Fig. 69 should make the situation quite clear. Like the planets, the Moon has no light of its own, and shines only because it reflects the light of the Sun; it is not even a very efficient 'mirror', since on average it reflects only about 7 per cent of the sunlight which falls upon it, and it would take around 465,000 full moons to give us

Planetary Data

Planet	Distance from Sun (millions of km)			Sidereal period	Rotation period	Axial inclination	Orbital inclination	Orbital eccentricity
	mean	max	min					
Mercury	57.9	69.7	45.9	88 days	58.6 days	Low	7°	0.206
Venus	108.2	109	107.4	227.7 days	243.2 days	178°	3°24'	0.007
Earth	149.6	152	147	365.3 days	23h 56m 4s	23°27'	0	0.017
Mars	227.9	249	207	687 days	24h 37m 23s	23°59'	1°51'	0.093
Jupiter	778	816	741	11.9 years	9h 50m 30s	3°4'	1°18'	0.048
Saturn	1427	1507	1347	29.5 years	10h 38m	26°44'	2°29'	0.056
Uranus	2870	3004	2735	84.0 years	17h 14m	98°	0°48'	0.047
Neptune	4497	4537	4456	164.8 years	16h 3m	28°48'	1°45'	0.009
Pluto	5900	7375	4425	247.7 years	6d 9h	118°	17°12'	0.248

	Equatorial diameter (km)	Density water =1	Mass Earth =1	Volume Earth =1	Escape velocity (km/s)	Maximum magnitude
Mercury	4,879	5.5	0.055	0.056	4.3	−1.9
Venus	12,104	5.25	0.815	0.86	10.4	−4.4
Earth	12,756	5.5	1	1	11.2	−
Mars	6,794	3.9	0.107	0.15	5.0	−2.8
Jupiter	143,884	1.3	318	1319	60.2	−2.6
Saturn	108,728	0.7	95	744	32.3	−0.3
Uranus	50,724	1.3	14.6	67	22.5	+5.6
Neptune	50,538	1.8	17.2	57	23.9	+7.7
Pluto	2,324	2.0	0.0022	0.006	1.18	+14

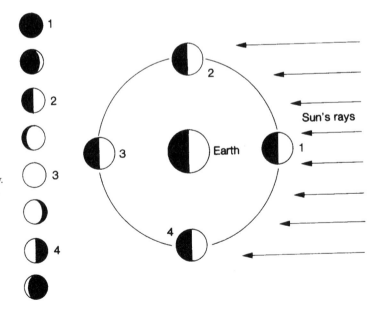

69. Phases of the Moon. 1. New. 2. Half (First Quarter). 3. Full. 4. Half (Last Quarter). (Not to scale.) The series to the left shows the appearance as seen from Earth.

as much light as we receive from the Sun.

In position 1 in Fig. 69, the dark side is turned towards us (new moon), and we cannot see the Moon at all – unless it passes straight in front of the Sun, blotting out the Sun briefly and causing a solar eclipse. Between positions 1 and 2 the Moon appears as a steadily-thickening crescent, and at position 2 it shows us half its sunlit side; this is termed First Quarter, because the Moon has completed one-quarter of its orbit. Between positions 2 and 3 the Moon is gibbous – that is to say, more than half but less than full. At 3, the Moon is full; between 3 and 4, gibbous; at 4, half again (Last Quarter), after which it again becomes a crescent, narrowing until it has arrived back at position 1. Though the Moon takes 27.3 days to go round the Earth, the interval between two new moons is 29.5 days, because the Earth is itself moving round the Sun.

Now let us turn to the planets. The so-called *inferior planets*, Mercury and Venus, have their own way of behaving, simply because they are closer to the sun than we are. Like the Moon, they show phases, and for much the same reason (Fig. 70), even though they move round the Sun and not round the Earth. When the planet is in position 1, it is new; at position 3, it is full; at positions 2 and 4 it is a half-disk, and is said to be at *dichotomy*. (In this diagram we have ignored the movement of the Earth itself round the Sun, because it makes no

difference to the basic principle.) At dichotomy, the planet is at its greatest *elongation* or angular distance from the Sun – 47° for Venus, but only 28° in the case of Mercury. At full (position 3) the planet is to all intents and purposes behind the Sun, so that it is above the horizon only during daylight, and is practically unobservable even with an accurately-aimed telescope. It is also at its greatest distance from the Earth, so that its apparent diameter is least (Fig. 71). When the planet is closest to us (position 1) it is new, and its dark side faces us, so that again it cannot be seen unless the lining-up is exact, when it may be seen as a dark spot in *transit* across the disk of the Sun. Because the orbits of both Mercury and Venus are tilted, transits are not common. The last transit of Mercury was in November 1999, and the next will be in May 2003; the last transit of Venus was that of 1882, and the next will not be until 2004. There can be nobody now living who can remember seeing a transit of Venus.

When an inferior planet is new, it is said to be at *inferior conjunction*; when full, it is at *superior conjunction*. Obviously, both Mercury and Venus always stay in the same general area of the sky as the Sun, so that within the naked eye they are visible only in the west after sunset or in the east before sunrise. Venus is relatively large and close, so that it is very bright and may even cast a shadow, but Mercury is much

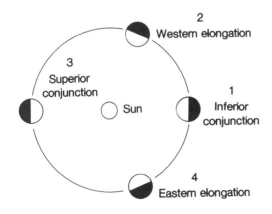

Earth

70. Phases of Mercury and Venus. 1. New; the planet cannot be seen (unless it crosses the disk of the Sun in transit). 2. Half (dichotomy). 3. Full. 4. Half again (dichotomy). Between 1 and 2, and 4 and 1, the planet is a crescent; between 2 and 3, and 3 and 4, it is gibbous.

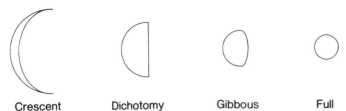

Crescent   Dichotomy   Gibbous   Full

71. Phases of Mercury and Venus. As the phase increases, the apparent diameter becomes less; when full, the planet is at its greatest distance from the Earth (superior conjunction).

more elusive, and there must be many people who have never seen it at all.

Things are very different with the *superior planets* – that is to say, Mars and beyond. For obvious reasons, they can never come to inferior conjunction. The orbit of Mars is shown in Fig. 72, but is fairly typical of all the rest.

When the Earth is at position E1 and Mars is at M1, Mars and the Sun are opposite in the sky; Mars is then said to be at *opposition*, and is best placed for observation, since it must be at its highest at midnight and is above the horizon throughout the hours of darkness. One year later the Earth has returned to E1, but Mars, moving more slowly in a larger orbit, has only reached M2. The Earth has to catch it up, so to speak, and this does not happen until the Earth is at E2 and Mars at M3, so that there is another opposition. The *synodic period*, or mean interval between successive oppositions, is 780 days, so that Mars is well placed only in alternate years. Thus there were oppositions in 1997 and 1999, but not in 1996 or 1998.

Mars has an orbit which is rather less circular than that of the Earth, so that not all oppositions are equally favourable. The opposition distance is not constant. It is least when Mars is also near perihelion, as in 1988, when the closest approach to Earth was just over 58,000,000 km; if opposition occurs when Mars is near aphelion, the closest approach may be no nearer than around 100,000,000 km.

Mars can pass through superior conjunction, but it is then on the far side of the Sun, and is out of view. It can never appear as a crescent or half, as seen from Earth, but when well away from opposition it

shows a distinctly gibbous phase, similar to that of the Moon a day or two from full.

There are times when Mars seems to move in a 'backwards' or *retrograde* direction against the stars for a while. Fig. 73 explains why. The Earth has an average velocity in orbit of 29.7 km per second, Mars only just over 24 km per second, so that Mars seems to retrograde as the Earth 'passes it on the inside'.

Obviously, Mars is brightest when closest to us, and on occasions (as in September 1988) it may actually outshine all the other planets apart from Venus. At its faintest it may sink almost to the second magnitude, not much brighter than Polaris, and it is then only too easy to confuse with a star.

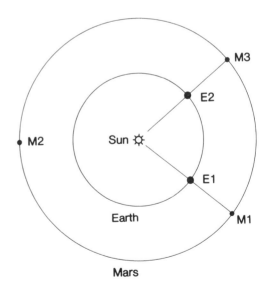

72. Orbit of Mars. Oppositions are not equally favourable.

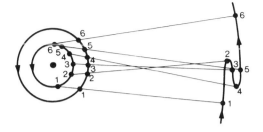

73. Apparent retrograde motion of Mars. As the Earth 'by-passes' Mars, by virtue of moving more quickly in a smaller orbit, Mars seems to move in a retrograde direction against the stars for a while before resuming its usual direct movement.

to opposition only a little over a day later every year.

All the planets and all known asteroids move round the Sun in the same sense as the Earth (*direct motion*), but this is not true of all comets, some of which move in retrograde orbits (note that this is not the same thing as the apparent retrograde movement of a superior planet against the stars as the Earth by-passes it). For example Halley's Comet, which has a very eccentric orbit taking it out well beyond Neptune, is retrograde (Fig. 74). Since a comet depends upon reflected sunlight, it is visible only when reasonably close to the Sun and the Earth, though it is true that it could be followed well after it had moved out beyond the orbit of Saturn. It has a period of 76 years; we will say more about it later. It was last at perihelion in 1986; we are afraid that most of the readers of this book will be rather elderly before they see it again, since the next perihelion passage is not due until the year 2061!

The other superior planets behave in the same way as Mars, but they cannot move so far before the Earth catches them up and by-passes them, so that their synodic periods are shorter. Jupiter's is 399 days, so that the planet is well seen for several months in each year; Saturn's is 378 days, and with Pluto the synodic period is only 366.7 days, so that it comes

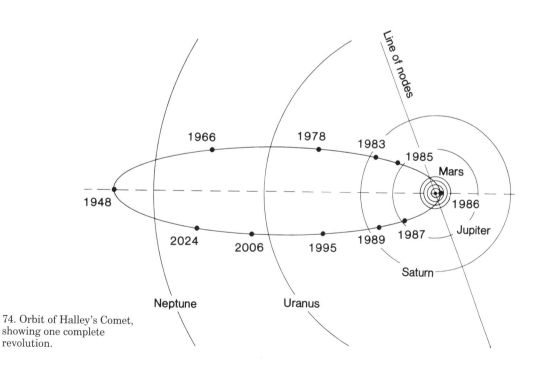

74. Orbit of Halley's Comet, showing one complete revolution.

Questions

1. (a) Why do the bright planets seem to keep to the Zodiac, instead of wandering about all over the sky?
   (b) Why do the giant planets contain more hydrogen and helium than the Earth does?
   (c) What do we believe to be the approximate age of the Earth and the other planets?
2. (a) Explain, with a diagram, why the Moon shows phases.
   (b) What is the phase of the Moon at First Quarter?
   (c) The Moon takes 27.3 days to complete one orbit, but the interval between successive new moons is over 29 days. Why?
3. (a) Why does Mercury show obvious phases, whereas Jupiter does not?
   (b) Are there any occasions when we can see Mercury when it is at inferior conjunction? If so, when?
   (c) Name the two inferior planets.
4. (a) Draw a diagram to show the orbits of the Earth and Mars. Indicate the positions of the two planets when Mars is at opposition.
   (b) Why are not all oppositions of Mars equally favourable?
   (c) Why are there times when Mars seems to have apparent retrograde motion against the stars?
5. (a) Explain why the synodic period of Saturn is shorter than that of Mars.
   (b) Why does Saturn appear less brilliant than Jupiter in the sky? Give two reasons.
   (c) Halley's Comet has an eccentric orbit, and has retrograde motion. Draw a diagram to illustrate this, putting in the orbits of the Earth and Neptune.

Practical work

1. Observe and draw the Moon's phases over a period of one lunar month, recording the dates, times and positions of rising and setting.
2. Plot the positions of selected planets over a period of time, putting in the stellar background.
3. Construct a simple moving model or *orrery* of the Solar System, showing the principal planets. (You need not keep to an exact scale.)
4. Construct a model to explain why the Moon shows its phases.
5. Using plasticine, or some such substance, make globes to show the relative sizes of the planets.

14
The Sun

As we have seen, the Sun is a star. There is nothing remarkable about it; it is neither particularly large nor particularly small by stellar standards, and its luminosity is by no means exceptional. Astronomers class it as a dwarf star, though it is only fair to add that dwarf stars are much more common than giants.

We will have more to say about the Sun's position in the Galaxy later. Meanwhile, it is enough to note that it is moving round the galactic centre; the revolution period of around 225,000,000 years is often called the 'cosmic year'. One cosmic year ago, the most advanced life-forms on Earth were amphibians; even the

dinosaurs lay in the future. It is interesting to guess as to what conditions here may be like one cosmic year hence.

The figure given for 'surface gravity' is in some ways rather meaningless, because you certainly could not stand on the surface of the Sun. Quite apart from the fact that it would be rather warm, the surface is not solid – indeed the Sun is made up entirely of gas. The average density is less than 1½ times that of water, though the outer layers are much more rarefied and the core region much denser than this value.

To discover what the Sun is really like, our first step must be to find out what

---

Data for the Sun

Mean distance from Earth	150,000,000 km (in round numbers)
Distance from centre of Galaxy	about 27,000 light-years
Time taken to revolve once round centre of Galaxy	about 225,000,000 years
Velocity round centre of Galaxy	19.7 km/sec
Apparent diameter	mean, 32'01"
Equatorial diameter	1,392,000 km
Density (water = 1)	1.41
Mass (Earth = 1)	332,946 (this is equal to 2×10^{27} tons, or 99 per cent of the entire mass of the Solar System)
Volume (Earth = 1)	1,303,600
Surface gravity (Earth = 1)	28
Escape velocity	618 km/sec
Surface temperature	5500°C
Core temperature	at least 14,000,000°C
Rotation period, mean	25.4 days
Time taken for light from the Sun to reach the Earth	8.3 minutes

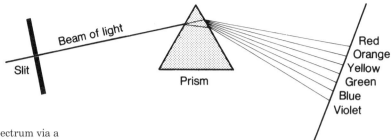

75. Production of a solar spectrum via a
slit and prism.

elements exist there, and this is where the spectroscope comes in. Newton passed sunlight through a glass prism and obtained a sort of rainbow effect, but it was not until the nineteenth century that much more was done. An English doctor, W.H. Wollaston, made some pioneer observations, but the story of solar spectroscopy really began in 1814 with the work of a German optician, Josef von Fraunhofer.

Fraunhofer passed sunlight through a slit, and then through a prism (Fig. 75). He built what was in fact a proper astronomical spectroscope, and found that the Sun's spectrum was made up of a rainbow band, from red at one end through to violet at the other, crossed by dark lines which are still often called Fraunhofer lines (see p. 76). He also found that the lines did not change; they were always in the same positions, and always had the same intensities. For example, there were always two prominent dark lines side by side in the yellow part of the spectrum. Altogether, Fraunhofer recorded over 570 different lines, and measured the position of 324 of them.

Unfortunately, Fraunhofer died young, and it was not until 1859 that the cause of the dark lines was worked out – mainly by Gustav Kirchhoff. Kirchhoff established that a luminous solid, liquid or high-pressure gas will give a rainbow or *continuous* spectrum, whereas a gas at low pressure will give an *emission* spectrum made up of isolated bright lines. Each element, or each combination of elements, produces its own special set of lines, making up a trade-mark which is strictly copyright. For example, the element sodium produces two bright yellow lines side by side (as well as many others); when these two yellow lines are seen, they must be due to sodium and nothing else.

Now let us look at the make-up of the Sun. The bright surface which we see is called the *photosphere*, and has a temperature of about 5500°C. It is composed of gas at relatively high pressure, and so it produces a continuous or rainbow spectrum. Above the photosphere is a layer of much thinner gas, the *chromosphere* (Fig. 76). This would be expected to give an emission spectrum of separate bright lines, but in fact the gases in the chromosphere are capable of absorbing those wavelengths which they would normally emit. Therefore the lines are *reversed*, and show up as the dark Fraunhofer or absorption lines crossing the background rainbow (Fig. 67). The result is a dark-line or *absorption spectrum*.

The vital point is that the positions and intensities of the lines are not altered. Our two dark lines in the yellow part of the solar spectrum correspond exactly to the two bright yellow lines of sodium, so that we can prove that there is sodium in the Sun. By now, over seventy of the 92 naturally-occurring elements have been identified there. One, the light gas helium, was found in the Sun before being tracked down on Earth.

Analysis of the Sun by means of the spectroscope shows that the most plentiful elements there are the two lightest, hydrogen (71 per cent) and helium (27 per cent), which does not leave much over for anything else. Hydrogen, which is much the commonest substance in the entire universe, is the main 'fuel' which keeps the Sun shining. It is not correct to say simply that the Sun is burning; if we had a Sun made up entirely of coal, burning as fiercely as the real Sun actually does, it would soon turn to ashes – but we know that the Earth is around 4.6 thousand million years old, and the Sun is certainly older than that, which means that we must look round for another source of energy. We find it in 'nuclear reactions'. But before going into this, we must pause to say something about the way in which matter is built up.

Matter is made up of *atoms*, each of which is almost inconceivably small. In the centre of each atom there is a *nucleus*, around which move less massive particles which we call *electrons*. This may sound straightforward enough, but in fact it is not, because it is misleading to picture either an electron or a nucleus as a solid lump. Things are much more complicated than might be expected, but we do not propose to go into detail here, because it does not come into the GCSE Astronomy syllabus. For the moment, we must be content with the easy-to-picture model of an atom as a kind of dwarf Solar System, with the nucleus taking the place of the Sun and the circling electrons representing the planets. It is an inadequate comparison, and it must not be taken literally, but it is the best we can do.

The nucleus of an atom is not a single particle, except in the case of hydrogen, whose atoms are the simplest of all (and even then there are different kinds of hydrogen). A nucleus contains *protons*, each of which carries a unit charge of positive electricity; it also contains *neutrons*, which have no net electrical charge at all.

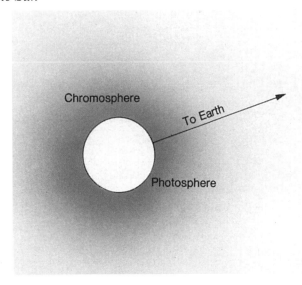

76. Production of the solar spectrum. The photosphere yields a continuous or rainbow spectrum. The chromosphere yields a line spectrum, and the dark absorption lines are superimposed on the rainbow background due to the photosphere.

An electron is much less massive than a proton, but its charge is equal in amount though opposite in sign. In a complete atom, the nucleus carries a positive charge which is exactly cancelled out by the combined negative charges of the planetary electrons, making the atom as a whole electrically neutral. For instance, an atom of helium has a nucleus with a positive charge of 2 units, because it contains 2 protons (together with 2 neutrons). To make up for this, there are two circling electrons. + 2 − 2 = 0, so that the complete helium atom has no charge at all.

There are 92 different fundamental kinds of substances known to occur on Earth. These are the *elements*. The lightest is hydrogen, with one circling electron (atomic number = 1); helium has 2 electrons, lithium 3, and so on up to uranium, which has no less than 92. There is a complete sequence from 1 to 92; one cannot have half an electron, so we may be sure that no elements remain to be discovered. Artificially, we can make elements with higher atomic numbers than 92, but all of them are unstable, so that they may not occur naturally anywhere in the universe.

There is another important flaw in our comparison between an atom and the Solar System. The electrons moving round a nucleus are limited to certain definite orbits, each of which represents a particular energy-level. If an atom absorbs energy, it is 'excited', and an electron may jump from one permitted orbit to the next; it cannot go half-way – the jump must be carried out all at once, or not at all. This state of excitation will not last indefinitely. The electron will jump back to its former orbit, closer to the nucleus, and in doing so it will make the atom emit radiation *at a definite wavelength corresponding to that particular change in energy-level*. Because we know a great deal about energy-levels, we can predict which wavelengths will be emitted.

If an atom absorbs more energy than it can cope with, one or more of the electrons may be torn away from it altogether, and the atom is said to be *ionised*; it is in fact incomplete. Now, remember that a complete atom is electrically neutral, because the positive charge of the nucleus is exactly balanced by the total negative charge of the circling electrons. If one or more of the electrons is removed, the atom will take on an overall positive charge. When it recaptures as many electrons as it had previously lost, it will be complete once more, and will again be electrically neutral. The radiation emitted during these various changes in energy-levels need not, of course, be visible light. It may be in any part of the electromagnetic spectrum; from the very short to the very long.

Now let us come back to the Sun, and see what reactions are taking place there.

Deep inside the Sun, the pressure is tremendous and the temperature is very high – at least 14,000,000°C, and perhaps rather more (between 15,000,000 and 16,000,000°C). Hydrogen gives us the key. Near the Sun's core, or 'power-house', the hydrogen nuclei are combining to form nuclei of the next lightest element, helium. It takes four hydrogen nuclei to make one nucleus of helium, but in the process a little energy is released and a little mass is lost. It is this energy which keeps the Sun shining. The mass-loss amounts to 4,000,000 tons per second, so that the Sun now 'weighs' considerably less than it did when you started reading this page, but please do not be alarmed; the Sun is so massive that it will not change noticeably for several thousands of millions of years yet.

The process of hydrogen-into-helium is not as straightforward as it might appear, and is accomplished by a roundabout method known as the proton-proton cycle, but the result is clear: hydrogen is being turned into helium, and this is why the Sun radiates. Energy is created deep inside the Sun, and is conveyed to the surface by convection (there is some analogy here with a tank of water which is heated from the bottom; the hot water rises to the top, carrying the energy with it). This is quite different from the way in which the energy reaches us on Earth, which is by straightforward radiation. Eventually the supply of available hydrogen 'fuel' must run out, and the Sun will change, but since this brings us on to the whole question of stellar evolution we will defer it until later. All we need do here is to note that all ordinary stars are shining by the same sort of process as takes place deep inside the Sun.

There is detail to be seen on the Sun's surface with a telescope, but please bear one thing in mind: never, under any circumstances, look straight at the Sun through a telescope, or even binoculars, even when a dark filter is placed over the eyepiece. If you do, then you will damage your eyes permanently. We will say more about this below, but we make no apology for stressing it, because accidents have happened in the past. There is only one

77. The great sunspot of 1947 – the largest ever seen.

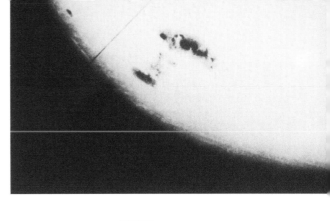

78. Sunspots – drawings which Patrick Moore made in July 1988 by projecting with a 3-inch (7.6-cm) refractor.

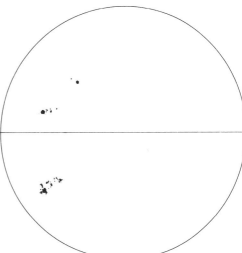

July 4, 1800 U.T.

July 5, 1200 U.T.

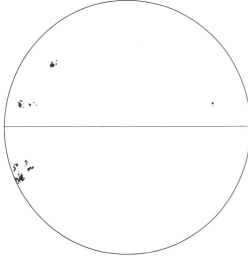

July 6, 1230 U.T.

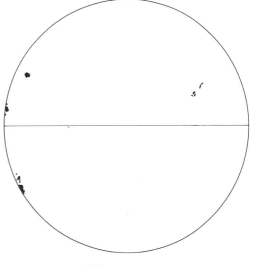

July 7, 0945 U.T.

golden rule about looking straight at the Sun with any optical equipment: *Don't*.

With indirect methods of observation, however (see p. 92), the view is interesting. In particular there are the sunspots, which are darker patches against the brilliant photosphere. The temperature of a spot is about 2000°C lower than that of the surrounding surface, which is why a spot looks dark; if it could be seen shining on its own, the surface brightness would be greater than that of an arc-lamp.

A large spot is made up of a darker central portion or *umbra*, surrounded by a lighter area or *penumbra*; the shapes are often very complicated, and there may be many umbrae contained in the same mass of penumbra. Perfectly circular spots do occur, but are the exception rather than the rule. Spots may also be isolated, but tend rather to appear in groups. A group often takes the form of two main spots, a leader and a follower, with various smaller spots around. The sizes are tremendous; the largest group ever recorded, that of April 1947, had an area of 18,130,000,000 square kilometres, and a large group seen in March 1989 was almost as large.

Since the Sun's surface is gaseous, we cannot expect the spots to be permanent, and we find that their lifetimes are limited. Really giant groups have been known to persist for more than six months, whereas very small spots may be born, develop, and disappear over a period of only a few hours. The shapes and forms change constantly, and the life-story of a group is fascinating to follow. Fig. 78 shows some drawings made in 1988; the changes from one day to the next are very marked.

Note, too, that the spots seem to be carried across the face of the Sun, from one side to the other. This is because the Sun is rotating. The rotation period is between 3 and 4 weeks, so that a spot will eventually be carried over the edge or limb of the disk, to reappear in a fortnight or so at the opposite limb – if, of course, it still exists. Some large groups may last for long enough to cross the disk several times.

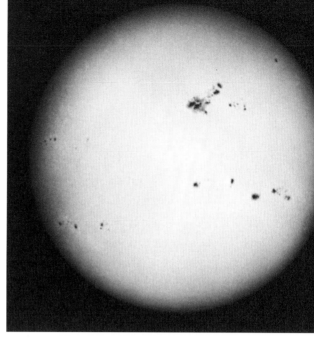

79. The face of the Sun, 13 June 1989, photographed by F. Dubois; 102 mm, f/15 refractor, exposure 1/1000 dec. on Agfa Ortho film. North is up.

Because the Sun is a globe, a regular spot will seem foreshortened when it is close to the limb, with the penumbra narrowest toward the centre of the disk. This effect – known as the *Wilson effect*, after the Scottish astronomer who first drew attention to it – indicates that most spots are hollows, not elevations.

Though sunspots have been known for so long, we have to admit that we are by no means certain exactly how and why they are produced. However, we do know that they are centres of strong magnetic fields, and that magnetic effects are closely involved in their origin. Note also that the Sun does not rotate in the way that a solid body would do; the rotation is shortest at the equator, and longest at the poles (where spots are never seen). The polar rotation period is around 35 days.

Active groups often produce *flares*, which are violent, short-lived outbursts – not generally visible in ordinary light; we have to use less direct methods of studying them. Flares emit charged particles and short-wave-length radiations, which cross the 150,000,000 km gap between the Sun and the Earth and cause rapid variations of the needles of magnetic compasses ('magnetic storms') and also overload the Van Allen belts, producing displays of aurorae. They also cause disturbances in radio and television reception, because they affect conditions in the upper part of the Earth's atmosphere which is known as the ionosphere. Reception is best when solar activity is at its lowest.

The Sun shows a roughly regular cycle of activity. Every eleven years or so it is particularly energetic, and there may be many spot-groups visible at the same time. Activity then dies down, and at solar minimum the disk may be free of spots for weeks at a time, but then the activity builds up once more toward the next maximum. The cycle is not completely regular, but it is very marked. Thus there were maxima in 1968, 1979-80 and 1990-91, with another due in 2000-01, and minima in 1976, 1987 and 1996-7. Not all maxima are equal in intensity, and there may be periods when the cycle is suspended altogether. So far as we can tell from the admittedly incomplete records, there were almost no spots between 1645 and 1715, and few aurorae were recorded; this is known generally as the Maunder Minimum (after the British astronomer E.W. Maunder, who pointed it out). Whether there is any direct connection between sunspot activity and the Earth's weather is by no means certain, but it is worth noting that the period of the Maunder Minimum was also marked by very cold conditions in Europe.

The positions of spot-groups on the Sun's disk are certainly affected by the cycle. At the start of a new cycle, spots tend to appear at high latitudes, but as the cycle progresses new groups break out at lower and lower latitudes. Before reaching the equator, however, the old-cycle spots die away, to be replaced by outbreaks of high-latitude groups of the next cycle. This behaviour was first noted by the German astronomer F.W. Spörer, and is therefore known as Spörer's Law. If the position of spots is plotted against time, the graph

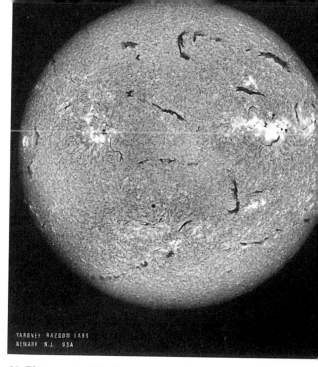

80. Photograph of the Sun taken in the H-Alpha Line with the Razdow solar optical telescope, WS250-1A Solar Observatory, Ramey Air Force Base, Puerto Rico, 15 July 1968.

produced is often called a 'butterfly diagram'.

Sunspot groups are often associated with active regions known as *plages*; the brightest features of this kind, known as *faculae* (Latin, 'torches') may be very conspicuous. Faculae often appear in areas where a spot-group is about to break out, and persist for some time after a group has vanished. Moreover, under good conditions it can be seen that no part of the solar surface is uniform; it has a granular structure. Each granule has a diameter of around 1000 km, and lasts for eight minutes or so. These granules represent upcurrents, and at any particular time it has been estimated that the whole surface contains about four million of them.

Next, we must come back to a well-worn theme: that of the danger of direct Sun-gazing. Even with the naked eye it is not wise to look straight at the Sun, and to use a telescope in this way is sheer madness. You will focus all the light and heat on to your eye, and you will blind yourself – probably permanently. This is true even when the Sun is low down and misty, and looks deceptively harmless. Even a second's glimpse may be disastrous. There are some devices which are fairly harmless, but dark caps or filters fitted over the eyepiece are emphatically not to be recommended, because they can never give proper protection, and in any case they are always liable to shatter without warning.

To observe sunspots, point the telescope towards the Sun by squinting along the top of the tube, or by casting the shadow of the telescope itself on to a screen held or fixed behind the eyepiece. The Sun's image will then fall on to the screen, and you will have an excellent view of any spots and faculae which may happen to be on the disk (Fig. 81). This is safe, but always take care to keep your eye well away from the eyepiece. Observing sunspots in this way makes one of the best of all GCSE projects, if for no other reason than that it can be done in the daytime. We have given some further details at the end of this chapter.

The Sun sends out radiations at all wavelengths; it is for example a source of radio waves – first detected by a British team during the war, and originally thought to be due to German jamming of our radar instruments. Studies of infra-red, ultra-violet and X-ray wavelengths have told us a great deal. There is also the *solar wind*, which is a continuous outflow of charged particles in all directions; it is made up of protons, helium nuclei ('alpha-particles') and electrons. The solar wind by-passes the Earth at a velocity of around 600 km per second, and the deep-space planetary probes have shown that it is still detectable at great distances, well beyond the orbits of Neptune and Pluto. It has a marked effect upon the tails of comets, as we will see later, and upon aurorae and the Van Allen belts.

During recent years we have found that the Sun is vibrating or oscillating; it has been said that the globe is 'quivering like a jelly'. But to go into more detail here would be beyond our scope, so let us now

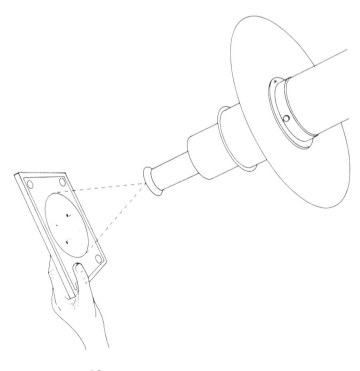

81. Projecting the Sun's image on to a screen behind the eyepiece.

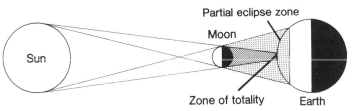

82. Theory of a solar eclipse. The Moon's shadow touches the Earth; to either side of the zone of totality, a partial eclipse is seen.

turn to the Sun's outer layers and atmosphere, which means dealing first with the spectacular phenomena known as solar eclipses.

The Moon moves round the Earth; the Earth moves round the Sun. Therefore, there must be times when the three bodies move into a direct line, with the Moon in the mid position. By a lucky coincidence (it is certainly nothing more), the Moon and the Sun appear virtually the same size in the sky; the Sun's diameter is 400 times greater than that of the Moon, but it is also 400 times further away. The result is that when the lining-up is exact, the Moon can just hide the bright surface of the Sun, so producing a *total solar eclipse* (Fig. 82). Because the Moon's shadow only just reaches the Earth, a total eclipse is visible only over a narrow belt on the Earth's surface. The track of totality can never be more than 272 km wide, and is usually less, so that one has to be in the right place at the right time. Moreover, no eclipse can be total for more than 7½ minutes. The maximum length of totality for the eclipse of 22 July 1990, for instance – visible from Finland and parts of Russia – was no more than 2 minutes 33 seconds.

A total eclipse of the Sun is a magnificent sight. As the last fraction of the photosphere is hidden by the advancing Moon, the sunlight fades; there is the 'diamond ring' effect just before the photosphere is covered, and then the Sun's outer atmosphere flashes into view. There are the masses of red hydrogen gas known as *prominences* (once, misleadingly, called Red Flames); there is the *chromosphere* itself, the lower part of which is sometimes called the 'reversing layer' because it is responsible for the dark Fraunhofer lines, and there is the pearly *corona*, which stretches for many millions of kilometres into space (Fig. 85). As soon as the first segment of the photosphere reappears, the prominences, the chromosphere and the corona vanish from sight. During totality, the sky becomes dark enough for planets and bright stars to be seen.

Instruments based upon the principle of the spectroscope make it possible to study the prominences and the chromosphere at any time, without waiting for an eclipse, but before the development of space research methods there were many investigations which could be carried out only during the fleeting moments of totality, so that

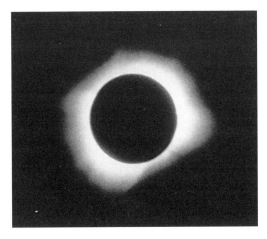

83. Total solar eclipse, 11 August 1999. Photograph by Chris Doherty.

93

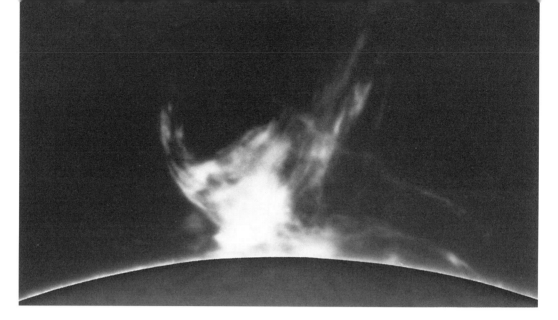

84. Active prominence rising from the Sun. This picture was taken with a special filter, not during a total eclipse.

astronomers were always ready to go on long journeys to take advantage of their opportunities – often, alas, to be disappointed when clouds covered the sky at the critical moment.

Prominences lie in the chromosphere, and may be very conspicuous. Quiescent prominences may hang above the Sun for weeks, though eruptive prominences change so rapidly that time-lapse photography has made it possible to produce superb films of them. They were first described by the Swedish observer Vassenius at the total eclipse of 1733, but it was not until the eclipse of 1842 that astronomers became quite certain that they belonged to the Sun rather than being due to an atmosphere around the Moon.

During totality, the corona dominates the scene. It is made up of gas, with a density less than one million millionth of that of the Earth's air at sea-level, but it is at a very high temperature of nearly 2,000,000°C. From this, you might think that it sends us a great deal of heat, but this is not so. The scientific definition of temperature is not the same as what we ordinarily mean by heat; it depends upon the velocities at which the various particles are moving around –

85. Above the Sun's bright surface or photosphere lies the chromosphere; beyond this, the corona.

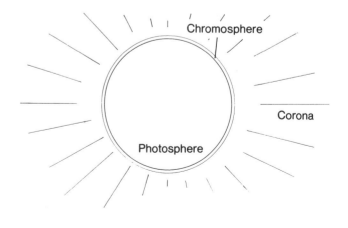

Chromosphere

Corona

Photosphere

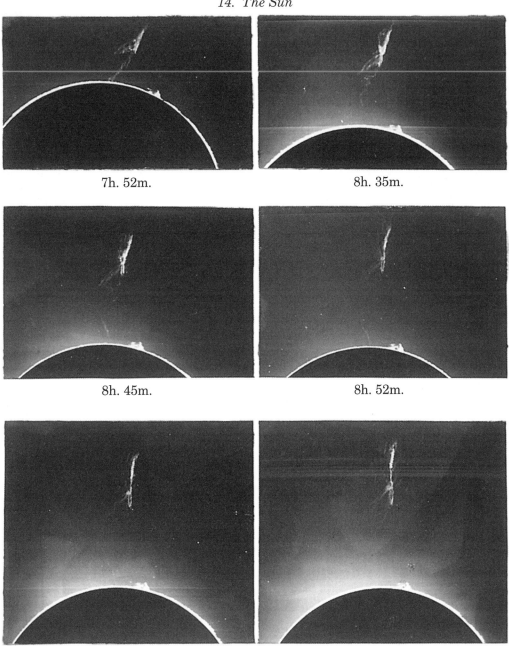

7h. 52m.

8h. 35m.

8h. 45m.

8h. 52m.

8h. 52m.

9h. 3m.

86. Solar prominences. These features are visible
with the naked eye only during a total eclipse of
the Sun, but with suitable equipment they can be
studied at any time. These famous pictures of an
eruptive prominence, reaching over 900,000 km
above the bright surface, were taken at Kodaikanal
Observatory on 19 November 1928.

87. Annular eclipse of the Sun, 29 April 1976, as photographed from Thera in the Greek islands. *Left*: Partial phase. *Right*: Annularity; a ring of sunlight is left showing around the dark disk of the Moon. An ordinary camera with a 400 mm telephoto lens, undriven, was used.

the greater the speeds, the higher the temperatures. In the corona, the speeds are tremendous, and the temperature is high, but there are so few particles that the amount of warmth is very slight indeed. The best way to explain this is to compare a firework sparkler with a red-hot poker. Each spark of the firework is at a high temperature, but it has so little mass that it cannot burn you – whereas one would be very reluctant to grasp the end of a glowing poker, even though its temperature is much lower.

The shape of the corona is affected by the sunspot cycle. Near solar maximum, the corona is usually fairly symmetrical; near spot-minimum it is less regular with streamers stretching across the sky. There are also areas in it where the pressures and temperatures are lower than average; these are known as *coronal holes*.

It is a pity that total solar eclipses are so rare as seen from any particular point on the Earth's surface. The last to be seen from any part of England was that of 11 August 1999, when the track crossed Cornwall. If you missed it, your next chance of seeing a total eclipse from England is in 2090!

If the Moon does not cover the Sun completely, the eclipse is *partial*, and the chromosphere, prominences and corona cannot be seen. The partial phase will be seen to either side of the track of totality, though there are many partial eclipses which are not total anywhere. There is also a third class of solar eclipse, the *annular* (Latin 'annulus', a ring). Because the Moon's orbit is elliptical, the distance between the Moon and the Earth varies considerably. When the Moon is near apogee – that is to say, its furthest from us – it appears smaller than the Sun, so that if it moves into the central position a ring of sunlight is left showing round the dark disk of the Moon (Fig. 88). Again, the phenomena of totality are not on view.

You may well ask why a solar eclipse does not happen every month. The answer is that the lunar orbit is tilted to ours at an angle of just over 5°, so that usually there is no eclipse; the Moon, new and therefore invisible because its unlit side is turned toward us, passes unseen either above or below the Sun in the sky.

By now we have found out a great deal about the Sun, but there is still much that we do not know. For example, theory predicts that the Sun should send out floods of strange particles called *neutrinos*, which have no electrical charge and no mass, so that they are difficult to detect. So far as we can tell, there are fewer neutrinos than there ought to be, and we do not know why.

What is the future of the Sun? It cannot go on shining for ever; eventually it must die, and the Earth itself will cease to

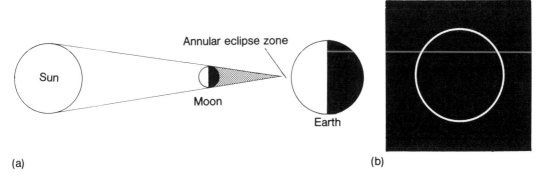

(a)

(b)

88. Annular eclipse of the Sun. (a) Theory: the main cone of shadow does not quite touch the Earth. (b) Appearance: a ring of the Sun is left showing round the dark disk of the Moon.

exist. But there is no immediate danger, and we are sure that the Sun will remain much as it is today for several thousands of millions of years yet. It may be nothing more than a dwarf star, but to us it is all-important, and it is hardly surprising that ancient peoples believed it to be a god.

Questions

1. (a) What is the approximate temperature of the surface of the Sun?
   (b) The corona is at a very high temperature, but produces little 'heat'. Explain this.
   (c) What are the Fraunhofer lines?
2. What is a sunspot? Describe the appearance of a typical spot-group, and explain why no spot can be kept in view for more than two weeks consecutively.
3. Write briefly about (a) the corona, (b) prominences, (c) magnetic storms, (d) the chromosphere, (e) faculae.
4. (a) Describe how you would set out to observe sunspots. Explain the precautions which must be taken, and give your reasons for them.
   (b) How are the dark lines in the solar spectrum produced?
   (c) Why were there more sunspot groups visible in 1980 and in 1989 than in 1985 and 1986?
5. (a) What is a total solar eclipse? Explain, with the aid of a diagram, why a total eclipse occurs. Why is it not visible over a complete hemisphere of the Earth? Why do astronomers regard total solar eclipses as important?
   (b) What are the other two kinds of solar eclipses? (Use diagrams.)
   (c) Why does not a solar eclipse occur every month?
6. (a) Why does a sunspot look dark?
   (b) During a solar eclipse, what is the phase of the Moon?
   (c) What effects does a solar flare have upon the Earth?

Practical work: observing sunspots

As we have seen, the only sensible and safe way to look at sunspots is to project the solar

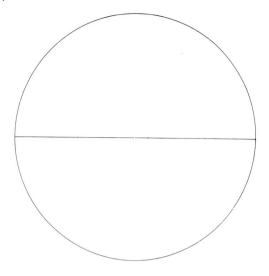

89. Solar disk with a diameter drawn across it so that the drawing can be properly orientated.

image on to a screen held or fixed behind the eyepiece of the telescope (for this, a refractor is better than a reflector, since the Sun's heat does the flat mirror of a Newtonian reflector no good at all). If your telescope is of more than 7.5-cm aperture, it is wise to stop down.

Draw a circle on a piece of paper; a convenient size is 15 cm, though with a very small telescope a 10-cm circle will do. Next, draw a diameter across the disk (Fig. 89). Alter the position of the screen until the focused disk of the Sun exactly fills the circle. (If you are simply holding the screen this is easy; if you are more ambitious, and have made a projecting box, you will have to make sure that it can be adjusted.) If there are any sunspots on view, allow the Sun to drift across the screen, and turn the screen so that the sunspots are running parallel with the diameter-line; this will ensure that the drawing is properly oriented. Then draw in the details of the spots as accurately as you can. If you make an observation on every clear day, you will soon see how the spots change, both in form and in position due to the Sun's axial rotation.

You will also see that circular spots are foreshortened when near the limb; the penumbra is usually narrower towards the centre of the disk (Wilson effect). This indicates that the spot is a hollow rather than a hump. Not all spots show it, but it is worth looking for. Look also for any bright faculae.

It is useful to fix a 'shield' over the telescope tube, as shown in Fig. 90; this will cast a shadow on to the screen, and make the details much easier to see.

You can also work out what is known as the Zürich or Wolf number, which is a measure of the Sun's activity. The formula is: $Z = k (10g + n)$, where Z is the Zürich number, k is a constant, g is the number of groups, and n is the total number of individual spots seen. The constant k depends upon the observer's experience and equipment, but for GCSE purposes it can be taken as 1 – so that to find the Zürich number, simply multiply the number of separate groups by ten and then add the number of individual spots seen.

More experienced observers will find plenty of interesting phenomena, but more detailed work requires more equipment than an ordinary telescope and eyepiece, and to go into it here would be beyond our scope.

Solar eclipses can be watched; time the moments when the Moon's limb covers and

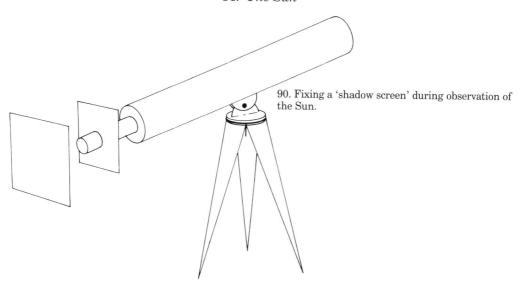

90. Fixing a 'shadow screen' during observation of the Sun.

then uncovers any visible spots. If you are fortunate enough to see a total eclipse, then look out for the chromosphere, prominences and corona; see whether any stars or planets come into view, measure the drop in temperature, and also look for 'shadow bands', which are wavy lines seen just before and just after totality. These are best seen against a white screen. They are, of course, phenomena of the Earth's atmosphere, and are surprisingly difficult to record.

Solar photography consists mainly of taking pictures of the screen upon which the disk is projected – a rather awkward process, because it is difficult to get the camera in the right position without obtaining a distorted view, but it can be done. Of course, proper solar photography is a different matter, and needs special equipment.

Finally: never forget that the Sun is a dangerous body. Treat it with due respect.

15
The Moon

The Moon is our companion in space. Officially it is regarded as the Earth's satellite – a secondary body – but this may not be entirely justified. There are several satellites in the Solar System which are comparable with our Moon: Ganymede, Callisto, Io and Europa in Jupiter's system, Titan in Saturn's, and Triton in Neptune's – but all these move round giant planets. Unless we count the exceptional Pluto-Charon pair, the Moon is much more massive *relative to its primary planet* than any other satellite. As we have seen, it has 1/81 of the mass of the Earth, whereas Ganymede, the largest satellite in the Solar System, has only 1/12820 of the mass of Jupiter. There are therefore grounds for regarding the Moon as a companion planet rather than as a satellite.

We have already described the phases of the Moon, from new (invisible) to full. There can be nobody who is not familiar with these, but two extra points seem worth noting. First, when the Moon is a crescent it is often possible to see the unlit side shining faintly; this is known to country folk as 'the Old Moon in the New Moon's arms', but officially as Earthshine, because it is due to light reflected on to the

Data for the Moon		
Distance from the Earth, centre to centre	max	406,697 km
	min	356,410 km
	mean	384,400 km
Distance from the Earth, surface to surface	max	398,581 km
	min	348,294 km
	mean	376,284 km
Orbital period	27.32 days	
Axial rotation period	27.32 days	
Synodic period (new moon to new moon)	29d 12h 44m	
Orbital inclination	5°09'	
Orbital eccentricity	0.054	
Mean orbital velocity	3680 km/h	
Diameter	3476 km	
Apparent diameter seen from Earth	max	33'31"
	mean	31'05"
	min	29'22"
Density (water = 1)	3.34	
Mass (Earth = 1)	0.012	
Escape velocity	2.38 km/s	
Surface gravity (Earth = 1)	0.165	

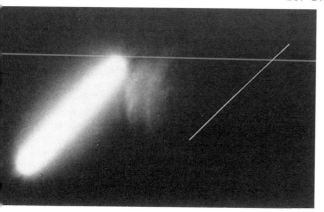

91. Trails of the Moon and Venus (R. Aylott). These were taken with a fixed camera, and show the drift across the sky.

Moon from the Earth. Secondly, so far as we can tell, the phases of the Moon have no effect on our weather. There is no reason why they should; the Moon is not necessarily closest to us (perigee) when it is full.

Because the Moon is so close to us, it seems to move quite rapidly against the starry background, travelling from west to east among the constellations. This means, of course, that it rises later night by night. The time-lapse between rising times on successive nights is called the *retardation*, and may amount to as much as an hour. However, remember that the Moon's path is inclined to the ecliptic by only a few degrees, so that the retardation depends upon the angle which the ecliptic makes with the horizon – since the Moon shifts eastward by the same actual amount every day (or nearly so). In late September, the ecliptic makes its shallowest angle with the horizon (Fig. 92), and so the retardation is less than usual, sometimes no more than a quarter of an hour or so. This is known as Harvest Moon. No significance is attached to it, and neither is it correct to say (as many books do) that around 22 September the full moon rises at practically the same time for several evenings in succession.

Incidentally, Harvest Moon looks no larger than any other full moon, and it is equally wrong to claim that a full moon is larger when low in the sky than when riding high above the horizon. This is the celebrated 'Moon Illusion'. Note also that because the full moon must always be opposite to the Sun in the sky, winter full moons are higher up than those of summer, because in winter the Sun drops to a greater distance below the horizon.

When the full moon is exactly lined up with the Sun and the Earth, it passes into the Earth's shadow, and we see a *lunar eclipse* (Fig. 93). This does not happen every month, because of the tilt of the Moon's orbit; usually the full moon passes either above or below the cone of shadow cast by the Earth, and avoids eclipse. The main cone of shadow is known as the *umbra*; because the Sun is a disk, not a point source of light, there is an area of *penumbra* to either side of the main cone. (Note that the terms 'umbra' and 'penumbra' as used here have nothing to do with the umbra and penumbra of sunspots.)

Eclipses of the Moon may be either total

92. The Moon moves along the ecliptic by virtually the same distance every day (for instance, between positions A and B). The sharper angle between the ecliptic and horizon in September means, therefore, that the retardation is less than at other times. This produces Harvest Moon.

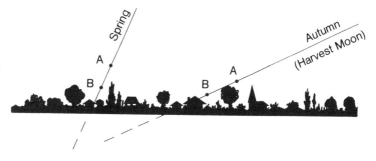

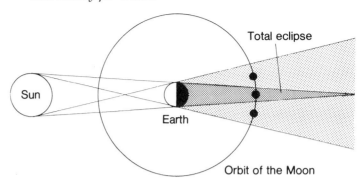

93. Theory of a lunar eclipse.

or partial. If the Moon does not move into the umbra, we see a penumbral eclipse, which is not too easy to detect with the naked eye; all you will notice is a very slight dimming. When a lunar eclipse is seen, it is visible from a complete hemisphere of the Earth – that is to say, from any point where the Moon is above the horizon at the time. This means that as seen from any given position on the Earth, lunar eclipses are more common than those of the Sun. Moreover, a lunar eclipse is a leisurely affair; totality can last for as long as 1 hour 44 minutes.

During a full eclipse, the supply of direct sunlight is cut off from the Moon's surface, but usually the Moon does not disappear, because some of the Sun's rays are bent or refracted on to it through the layer of atmosphere surrounding the Earth. All that happens is that the Moon turns a dim, often coppery colour. Lunar eclipses are not important, but they are interesting to watch, and on many occasions the colours are beautiful; everything depends upon the prevailing conditions in the Earth's atmosphere, and some eclipses are much 'darker' than others.

During its movements among the stars, the Moon sometimes passes directly in front of a star, and hides or *occults* it. When this happens, the star shines steadily right up to the moment when it is covered, when it snaps out as quickly as a candle-flame in the wind. This is because the Moon has no atmosphere. If there were

a layer of air round the lunar limb, the star would flicker and fade for a few seconds before disappearing (Fig. 94).

The lack of atmosphere is due to the Moon's low escape velocity. Even if there were an atmosphere thousands of millions of years ago, the weak lunar gravity could not hold it down, and the atmosphere leaked away into space. Today, the Moon is an airless world. There are no clouds, no weather, and no sound. The Moon is indeed a quiet place.

The Moon spins on its axis in exactly the same time that it takes to complete one orbit: 27.3 Earth-days, so that its rotation is said to be *captured* or *synchronous*. There is no mystery about this. Early in its history, it must have been spinning round more quickly; but it was not solid – it had not cooled down sufficiently after its formation – and so the Earth raised powerful tides in its globe, trying to keep a bulge turned Earthward. This slowed down the Moon's

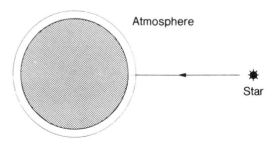

94. If the Moon had an atmosphere, the star's light would flicker and fade before occultation.

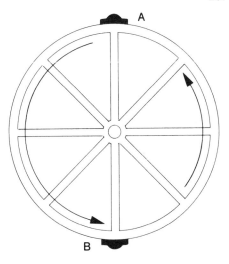

95. Rotating wheel being slowed down by opposite brake shoes.

constant rate; following Kepler's Laws, it moves quickest when closest to us (i.e. at perigee). This means that the rotation and the orbital position become periodically 'out of step', as it were, and the Moon seems to rock very slightly from side to side, so that we can see a little way beyond alternate edges. This is called *libration in longitude*. Of course, the rocking is too slow to be noticed over a short period, but it does have obvious effects. For instance, under favourite libration the well-formed grey plain called the Mare Crisium seems well on to the disk; under opposite libration its edge is not far from the lunar limb.

The cause of libration in longitude should be explained fully by Fig. 97. There is also a *libration in latitude*, because the Moon's equator is appreciably tilted to the plane of its orbit; and there is

rotation, rather in the manner of a cycle-wheel spinning between two brake shoes (Fig. 95). Eventually the rotation had been slowed down so much that relative to the Earth, it had stopped altogether, so that the Moon now keeps the same face turned permanently toward us. Note, however, that the Moon does not keep the same face turned permanently towards the Sun, so that every part of its surface has regular day and night. A lunar 'day' is almost as long as two Earth weeks.

To make this quite clear, picture a boy walking round a chair (Fig. 96), turning as he walks so as to keep his face turned toward the chair all the time. After one circuit he will have completed one rotation on its axis, and will have faced every wall of the room, but anyone sitting on the chair will never have seen the back of his neck. From Earth, we never see the 'back' or far side of the Moon, and before the Space Age we knew nothing definite about it.

However, there is a minor qualification. The Moon spins on its axis at a constant rate, but because its orbit is elliptical it does not move along at a

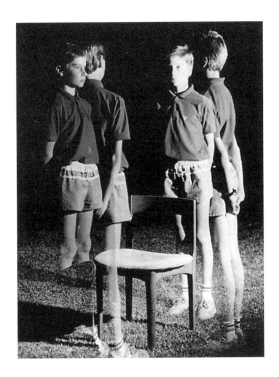

96. Chris Doherty, the boy walking round the chair, keeps his face turned toward the chair all the time – yet he faces every wall of the room in turn.

a *diurnal libration*, because the Earth is rotating and taking the observer with it. When the Moon is on the horizon, the observer is 'elevated' above the centre of the Earth by about 6400 km (the Earth's radius), and can see for an extra degree round the Moon's edge (Fig. 98).

This may sound rather confusing, but it is easily sorted out. All in all, the various librations mean that from Earth we can study a total of 59 per cent of the Moon's surface, though obviously we can never see more than 50 per cent at any one moment. The remaining 41 per cent is permanently out of view. To learn anything positive about it, we had to wait until 1959, when the Russians sent their rocket Luna 3 round the far side and sent

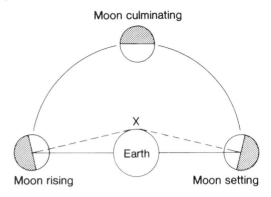

98. Diurnal libration. At moonrise, an observer at X can see a little way beyond the mean limb; at moonset he can see a little way beyond the other mean limb.

back the first photographs of it.*

The markings on the Moon are easy to see with the naked eye; binoculars will show them very clearly, and with any small telescope there is a tremendous amount of detail to be made out. In particular, there are the dark plains which are still known as seas even though there has never been any water in them (one cannot have water upon an airless world). They have been given attractive names, such as the Sea of Serenity (in Latin, Mare Serenitatis); the Ocean of Storms (Oceanus Procellarum); the Bay of Rainbow (Sinus Iridum) and so on. Astronomers always use the names in their Latin form, and we propose to do the same here. Mare, pronounced 'mah-ri', is Latin for 'sea'; plural maria ('mah-ri-a').

Because of the captured rotation, the seas always appear in the same positions on the disk apart from the relatively minor effects of libration. From the northern hemisphere, for example, the Mare Crisium (Sea of Crises) will be to

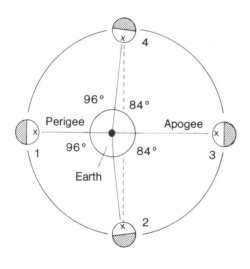

97. Libration in longitude. Take X as being the centre of the disk as seen from Earth, and begin with the Moon at perigee (position 1). After a quarter of its journey round the Earth, the Moon has reached position 2; but as it has travelled from perigee, it has moved slightly quicker than its mean rate, and has covered 96° instead of 90°. As seen from Earth, point X is slightly displaced from the apparent centre of the disk, and a small portion of the 'far side' has come into view. At position 3, X is again central. A further 84° is covered between positions 3 and 4, so that X is again displaced, but this time we see a little way round the other limb. At the end of one revolution, the Moon has returned to position 1, with X once more central.

* One is reminded of the poem written long ago by someone who was irritated at being unable to see the far side of the Moon:

O Moon, lovely Moon with the beautiful face,
Careering throughout the bound'ries of space,
Whenever I see you, I think in my mind
Shall I ever, O ever, behold thy behind?

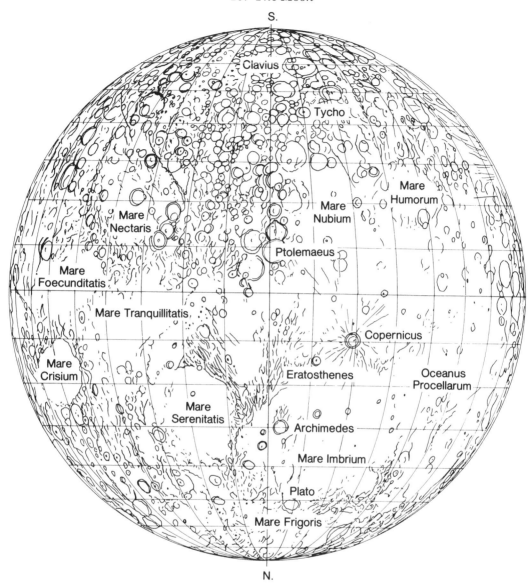

99. Outline chart of the Moon, showing the main maria and a few selected craters.

the upper right-hand side of the disk, the vast Mare Imbrium (Sea of Showers) to the left, rather above centre, and so on.

Most of the seas are joined together, though the Mare Crisium is separate. Some are roughly regular in outline, while others, such as the largest of all –

the Oceanus Procellarum or Ocean of Storms – are irregular. The main dark areas are shown in Fig. 99.

Some of the seas are bordered by mountain chains such as the Apennines and the Alps, which make up part of the boundary of the Mare Imbrium; the Mare Imbrium itself has a diameter of 1300 km, and is easy to identify even with the naked eye. Isolated peaks and hills are

100. The full moon, photographed by Peter Foley with his 30-cm reflector. Actually the Moon is just past full; the Mare Crisium (lower left) is crossed by the terminator. The dark patch to the right is the crater Grimaldi. Tycho, the great ray-crater, is near the top of the photograph (the picture is oriented with south at the top, as in the telescopic view), and the other major ray-crater, Copernicus, is slightly to the right of centre.

also very common. However, the entire lunar scene is dominated by the walled formations which are always known as craters. In size, they range from huge enclosures well over 250 km in diameter down to tiny pits. In general they have been named after famous men and women – usually, though not always, astronomers. Ptolemy, Kepler, Copernicus, Galileo and Newton, for example, all have craters named after them.

The profile of a typical large crater is shown in Fig. 101. Obviously it is nothing like a steep-sided pit or mine-shaft; the walls rise to only a modest height above the surrounding country, and the floor is sunken. With many craters there is a central mountain, or perhaps a group of peaks, which may be quite high even though they never reach the altitude of the surrounding wall. The walls themselves may be terraced and massive. Copernicus, on the Mare Nubium (Sea of Clouds) is a particularly magnificent crater, 97 km in diameter, with a central mountain group. On the other hand Plato, in the highlands between the Mare Imbrium and the Mare Frigoris (Sea of Cold) has no central mountain, and its floor is dark and relatively smooth, so that it is always easy to identify. Like Copernicus, it is 97 km across.

Like most craters, Plato is circular, but because it lies well away from the apparent centre of the disk it appears foreshortened. The closer to the limb, the greater the foreshortening, and in the 'libration' areas a crater looks like a very narrow ellipse. In fact, it is often difficult to tell the difference between a crater and a ridge. The effect is well shown by the Mare Crisium. It looks as though it is elongated north-south; in fact, the east-west diameter is greater (590 km, against 460 km).

When one crater breaks into another, it is almost always the smaller crater which breaks into the larger; the distribution is not random, and there are also many cases of crater-pairs, crater-groups and crater-chains. Note, for instance, the

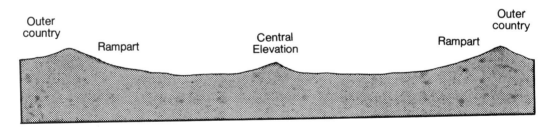

101. Profile of a typical large lunar crater.

106

102. The crater Tsiolkovskii, on the far side of the Moon, as photographed from Apollo 8 in 1968. Tsiolkovskii has a dark floor which seems to be a lake of solidified lava, and there is a pronounced central peak; in some ways the formation appears to be intermediate between a Mare and a crater. The only men who have had actual views of it are the crews of Apollos 8, 10, and 11 to 17!

splendid chain near the centre of the disk; there are three members – Ptolemaeus, Alphonsus and Arzachel – of which the largest, Ptolemaeus, has a diameter of no less than 148 km.

Some of the craters, notably Copernicus and the even more conspicuous Tycho, in the southern uplands, are the centres of systems of bright streaks or rays, which spread outward in all directions. They are surface deposits, and are not visible except when the Sun is fairly high over them; they are most conspicuous at full moon, when the Tycho and Copernicus rays dominate the entire scene. There are also many minor features, such as valleys, swellings or domes, and the crack-like rills, otherwise known as rilles or clefts.

A lunar crater is best seen when it is near the *terminator*, or boundary between the sunlit and night hemispheres of the Moon. The crater-floor is then filled with shadow, either totally or partially. As the Sun rises over it, the shadows shorten; near full moon there are almost no shadows at all, and the crater may be hard to identify unless it has a very dark floor (as with Plato) or has very brilliant walls (as with Aristarchus on the Oceanus Procellarum). Aristarchus, 37 km across and with a central peak, is the brightest crater on the entire Moon, and can often be seen when it is illuminated only by earthshine. Unwary observers have mistaken it for a volcano in eruption.

Full moon, then, is the very worst time to begin observing. To learn your way around the lunar surface, make drawings of the craters under different conditions of illumination. It is fascinating to follow the shadows as they change from one night to the next.

It was formerly believed that the lunar craters were of volcanic origin, and were related to terrestrial calderae, but it now seems certain that the craters are of impact origin. The sequence of events may have been as follows. The Moon was formed about 4600 million years ago. The heat generated during the formation made the outer layers molten down to a depth of several hundred kilometres; less dense materials separated out to the surface, and in the course of time produced a crust. Then, between 4400 and about 4000 million years ago, came what is termed the 'Great Bombardment', when meteorites rained

103. The lunar Sinus Iridum (Bay of Rainbows). The oval dark-floored crater to the left is Plato, 60 miles in diameter. Between Plato and Sinus Iridum, note the curious Straight Range. Photo by Commander H.R. Hatfield, 29-cm reflector.

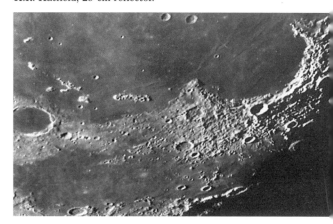

104. The lunar Mare Humorum, with the large crater Gassendi at its lower (northern) border. Taken in 1981 with a 39-cm reflector.

far side, and there are no major maria, though craters abound.

In general the lunar mountains form the boundaries of the regular 'seas'; thus the Alps and the Apennines make up parts of the border of the large Mare Imbrium. Isolated peaks and clumps of peaks are scattered all over the lunar surface.

The Moon today is virtually changeless, and even the 'youngest' craters such as Tycho are very old by terrestrial standards. Occasional local obscurations have been reported in some areas, and seem to be due to emissions from below the crust, but are very minor. Certainly no major formations have been produced for at least a thousand million years, and probably rather longer.

down to produce the oldest basins such as the Mare Tranquillitatis. The Imbrian basin dates back around 3850 million years, and as the Great Bombardment eased there was widespread vulcanism, with magma pouring out from below the crust and flooding the basins to produce such features as the complex, ringed Mare Orientale. Craters with dark floors, such as Plato, were also flooded at this time. The lava-flows ended rather suddenly, by cosmical standards, and for the last few thousand million years the Moon has seen little activity, apart from the formation of occasional impact craters such as Copernicus and Tycho. The ray systems centred upon a few craters are certainly late-comers, since they cross all other formations. We also have domes, gentle swellings of internal origin often with central craterlets; rills, which are cracklike features, sometimes associated with chains of craterlets; and wrinkle ridges, which cross some of the main seas such as the Mare Serenitatis.

There are differences between the Earth-turned and averted hemispheres of the Moon, since the crust is thicker on the

105. Lunar craters, photographed by Commander H.R. Hatfield (29-cm reflector) in 1966. South is to the top. The large lower crater is Ptolemaeus; centre, Alphonsus; upper, Arzachel.

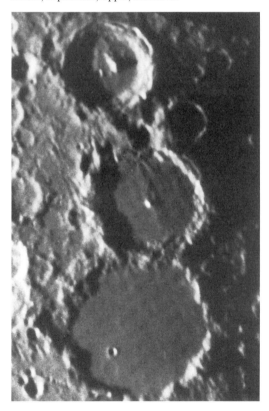

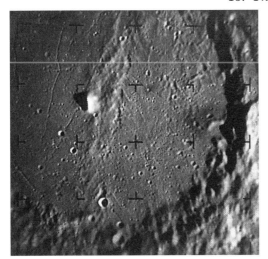

106. The floor of the crater Alphonsus, taken from Ranger 9 on 24 March 1965. The Rangers were probes which crash-landed on the Moon, sending back close-range pictures just before impact. The central peak of Alphonsus is well shown, as well as the rills and the craterlets.

The first successful space-probes to the Moon were sent up by the Russians in 1959. The third of these – Luna 3, launched on 4 October, exactly two years after Yuri Gagarin became the first man to orbit the Earth – went round the Moon and sent back the first pictures of the far side. As expected, the new regions were basically the same as those which we have always known. There were no really large 'seas', but there were craters, mountains, valleys and ray-systems. Other probes, both Russian and American, soon followed, and within a few years our knowledge of the Moon had been improved beyond all recognition.

One particularly important question was resolved on 31 January 1966, when the Soviet probe Luna 9 made a controlled landing in the Oceanus Procellarum. It had been suggested that the lunar surface might be covered with soft dust, kilometres deep, in which case sending an expedition there would have been very difficult indeed. Luna 9 showed that this was not so; it came down on a hard surface,

and did not sink. Then, in 1966 and 1967, the American Orbiters were put into closed paths round the Moon and sent back thousands of high-quality photographs, which made it possible to draw up detailed maps of the whole of the lunar surface. The stage was set for the Apollo missions.

Sending a manned space-ship to the Moon is no easy matter. To launch it, land it on the Moon and bring it directly back again would need more fuel than could possibly be carried. With the Apollo programme, the actual Lunar Module was taken with the main space-craft; once in orbit round the Moon, the Module was separated and made the final descent by using its own motor. Two of the three astronauts went down to the surface, while the other member of the crew remained in orbit. When the time came to depart, the lower part of the Module was used as a launching-pad; the two Moonwalkers returned to orbit, and rendezvoused with the main space-craft for the flight back to Earth.

107. Apollo 11; 'Buzz' Aldrin stands on the surface of the Moon, at Tranquillity Base.

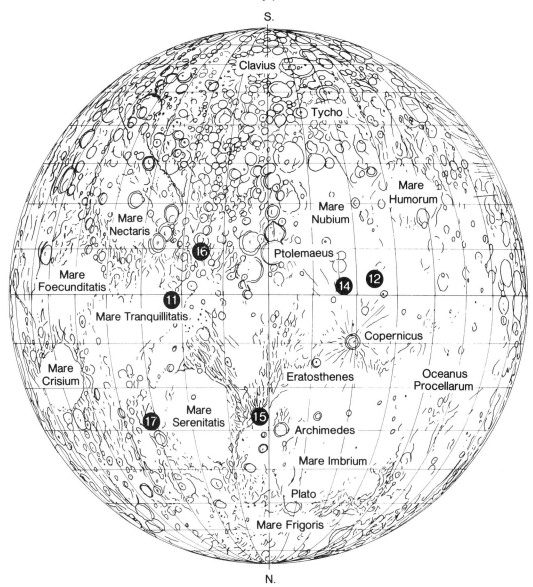

108. Landing sites of the Apollos:

	Lat.	Long.	Area
11	00°67'N	23°49'E	Mare Tranquillitatis
12	03°12'S	23°23'W	Oceanus Procellarum
14	03°40'S	17°28'W	Fra Mauro
15	26°06'N	03°39'E	Hadley-Apennines
116	08°60'S	15°31'E	Descartes
17	20°10'N	30°46'E	Taurus-Littrow

Two preliminary missions were dispatched, and then, on 20 July 1969, came the first actual landing. Apollo 11, carrying Neil Armstrong, Edwin (Buzz) Aldrin and Michael Collins, was put into lunar orbit; then Armstrong and Aldrin, in the module *Eagle*, made the descent. Armstrong's words, 'The *Eagle* has landed', were heard by television and radio up the American flag there was no breeze to make it flutter. As Armstrong and Aldrin moved around on the Moon, wearing their cumbersome space-suits, they seemed to walk in slow motion, because they had only one-sixth of their Earth weight. They stayed outside the *Eagle* for just over two hours, which was long enough to set up equipment capable of sending back signals long after the astronauts had left. They collected rocks, made measurements of various kinds, and also set up a seismometer to see whether there were any 'moonquakes'. When they were ready to depart, they fired the Eagle's ascent engine, and blasted back to rejoin Collins in orbit.

109. Varied surface features of the Moon are visible in this oblique view, looking north, from the orbiting Apollo 15 spacecraft. In foreground are the Hadley Rill and Apennine Mountains. David Scott and James Irwin landed just to the right of the rill's 'chicken beak'. At upper left is the crater Autolycus, diameter 40 km, and just beyond it the crater Aristillus, diameter 55 km. At upper right are the Caucasus Mountains and a portion of the Sea of Serenity. At lower left is the Marsh of Decay. This picture was taken with the 3-inch (75mm) mapping camera.

110. The Moon from Apollo 14, February 1971. The landing area was in the highlands of Fra Mauro. The picture shows the ALSEP (Apollo Lunar Surface Experimental Package) which handles power and data distribution; in the background are several of the scientific devices which had been set up by the two astronauts, Alan Shepard and Tom Mitchell, while part of the 'moon-cart' is shown in the left foreground.

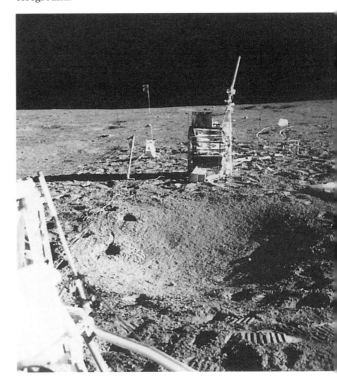

listeners all over the world. A few hours later, first Armstrong and then Aldrin stepped out on to the bleak rocks of the lunar Sea of Tranquillity.*

They found a strange scene. The sky was black, even though it was daytime; on the airless Moon, the sky is always black. There were hills and pits everywhere; nobody has ever bettered Aldrin's description of the view as 'magnificent desolation'. No sound, no movement; when the astronauts set

* I remember it very well. As an astronomer, I had been involved in the official mapping of the Moon; during the Apollo 11 mission I was broadcasting on BBC television. (P.M.)

111. Apollo 15; James Irwin with the Lunar Roving Vehicle.

Note that there was only one ascent engine. It had to work properly, first time; there could be no second chance. Luckily, all went well. The great breakthrough had been made.

Apollo 12 followed in November 1969. Apollo 13, in 1970, was nearly a disaster; on the outward trip there was an explosion in the service module, which carried the main propulsion unit, and it was only by a combination of skill, courage and luck that the astronauts returned safely to Earth. Then, between 1971 and 1972, came four more landings, on the last three of which the astronauts were able to drive around the Moon in 'Lunar Rovers' which they had brought with them. The Apollo programme ended in December 1972. Since then there have been no more manned expeditions, but there have been various unmanned missions. For example, several sample-and-return probes were sent up by the Soviet Union in the 1970s,

and the Clementine probe, sent up by NASA in 1994, produced very detailed maps of the surface. Radar had indicated that there might be ice inside some of the polar craters, whose floors are always in shadow and which therefore remain bitterly cold. In 1999 the lunar probe Prospector was deliberately crashed into a polar crater, in the hope that the cloud of débris thrown up would show traces of water. No such traces were found, and it does indeed seem that lunar ice is decidedly improbable.

The various expeditions have told us that the Moon's surface is covered with a loose layer or *regolith*, up to 20m deep in places. Below this comes more solid rock, and deeper still is the core, which is hot enough for the material to be molten. Moonquakes do occur, though they are very mild by our everyday standards and are certainly not violent enough to shake down any future Lunar Base. The rocks themselves are volcanic – mainly basalts – and in some cases are over four thousand million years old.

One other point was finally settled. The Moon showed no sign of life, either past or present. The lunar world has been sterile throughout its long history.

What will happen next? Plans for future missions are already being made. The Moon will be an ideal place for a scientific laboratory as well as an astronomical observatory, and if all goes well we may expect a fully-fledged Lunar Base to be established before the end of the century. Some of you now studying for GCSE may well go to the Moon, and see the black sky, the distant Earth, the barren craters and the lunar mountains for yourselves. If so, you will find plenty to interest you.

Questions

1. Describe the appearance of the Moon through binoculars or a low-power telescope. What are (a) the maria, (b) the ray-system, (c) the rills or clefts?

2. (a) Describe, with a diagram, the cause of a lunar eclipse.
   (b) What is the phase of the Moon during a lunar eclipse?
   (c) From any particular point on Earth, why are lunar eclipses more common than eclipses of the Sun?

3. (a) What is Harvest Moon, and when does it occur?
   (b) Why does the full moon appear higher in the sky at midsummer than in midwinter?
   (c) What is meant by 'retardation'?

4. (a) Why can we on Earth see only part of the total surface of the Moon?
   (b) Why has the Moon no atmosphere – what is one simple observational proof of this?
   (c) Why is the lunar sky always black?

5. (a) Who were the first men to land on the Moon, and what did they see?
   (b) Describe the constitution of the Moon (use a diagram if necessary).
   (c) Outline the main advances in our knowledge of the Moon obtained by the manned and unmanned probes.

Practical work

The outline map given overleaf (Fig. 112) will help in one of the best GCSE projects: observing and drawing craters and other lunar features under various conditions of illumination. On the map, south is at the top, as seen in an ordinary telescope; with the naked eye or binoculars, of course, north is at the top, as seen from the northern hemisphere of the Earth. Conventionally, Mare Crisium is to the *east*.

As we have seen, the boundary between the daylit and night hemispheres is known as the *terminator*. Instead of being hard and regular, the terminator is very rough and uneven, because the lunar surface is so mountainous. Most craters are best seen when close to the terminator, because their floors are then partly or wholly covered with shadow. Near full moon, the shadows almost disappear, and a crater such as Theophilus, which is striking when partly shadowed, becomes hard to identify. Moreover, the bright rays coming from Tycho, Copernicus and other centres (such as Kepler) dominate the scene near full phase, and tend to mask the other details. Of the craters given here, Plato, Grimaldi and Riccioli are always easy to locate, when in sunlight, because their floors are so dark; Aristarchus and Proclus are very brilliant. Note again that craters away from the centre of the apparent disk are foreshortened into long, narrow ellipses.

When drawing a lunar crater, do not attempt too large an area at one time. If the scale is too small, minor details are bound to be exaggerated and misplaced. Outlines of two prominent craters, Theophilus and Copernicus, are given in Fig. 112; they have been prepared from photographs. A good exercise is to trace these outlines on a white card, and then observe the actual craters through a telescope, filling in all the details you can see. Photographic outlines of other craters can be prepared in the same way. After a little practice you will find that you are 'learning your way around', and will be able to profit by using a more detailed map.

Sometimes the Moon passes in front of a star, and occults it. Timing these occultations is a useful project, because it gives the exact position of the Moon at that moment, but

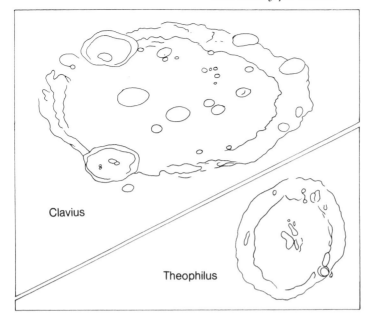

Clavius

Theophilus

112. Outlines of the craters Clavius and Theophilus.

great care is needed, and you must have an accurate stopwatch. Yearly lists of occultations are given in publications such as the Handbook of the British Astronomical Association.

If you have a camera with a telephoto lens, try to take pictures of the Moon, remembering that if you use an exposure of more than a second or two the Moon's motion will result in blurring of the image.

16

The Planets

As we have seen, the Solar System is divided into two definite parts. We have the four small inner planets, then a wide gap, and then the giants. It is now time to say something about the planets themselves. Before the Space Age, our knowledge was very limited; we have learned more during the past thirty years than we had been able to do during the previous thirty centuries – and we have had plenty of surprises.

Mercury is by no means easy to study from the Earth. It never comes much within 80 million kilometres of us, and it is always inconveniently close to the Sun in the sky. In fact, the best Earth-based observations of it are made when it is high above the horizon – that is to say, in broad daylight. But sweeping around for it is not to be recommended; there is always the danger of looking at the Sun by mistake. To locate it in daylight, you need a telescope equipped with accurate setting circles. Even when you find it, you will see nothing apart from the characteristic phase.

Using very large telescopes, some observers managed to record darkish patches and brighter areas, but that was all. The best pre-Space Age map was drawn by E.M. Antoniadi in 1934, with the help of the powerful 83-cm refractor at the Paris Observatory, but we now know that it was very inaccurate. Antoniadi believed, too, that the rotation was synchronous, and equal to Mercury's revolution period round the Sun (88 Earth days). In this case, one part of the surface would have been in constant daylight and another part in permanent night, with only a narrow 'twilight zone' in between over which the Sun would rise and set, always keeping close to the horizon. The low escape velocity meant that there could be nothing much in the way of atmosphere.

Then, in 1962, it was found that the dark side of the planet is much warmer than it would be if it never received any sunlight. The true rotation period is 58.6 days, or two-thirds of a Mercurian 'year'. It so happens that whenever Mercury is best placed for observation from Earth, the same face is turned toward us, and this is what misled Antoniadi.

Only one space-probe has so far flown past Mercury. This was Mariner 10, which made three active passes of the planet (March and September 1974, and March 1975) before it ceased transmitting. Excellent pictures were obtained

113. Mercury, from Mariner 10; a mosaic, taken as the space-probe drew away from the planet on its last active encounter in March 1975.

115

of almost half the total surface, and it was found that Mercury, like the Moon, is mountainous and cratered; there is one particularly striking feature, known as the Caloris Basin, which is 1300 km in diameter, and is bounded by a ring of smooth mountain blocks rising to almost 2 km above their surroundings. 'Caloris' indicates 'heat', and the Caloris Basin is one of the hottest places on Mercury, because when the planet is at perihelion the Sun is overhead there, sending the temperature up to over 450°C.

As expected, there is barely a trace of atmosphere, but there is a definite magnetic field – which is not surprising, because Mercury is as dense as the Earth, and must have a relatively large, iron-rich core. But it is hostile in every way, and manned expeditions there seem to be out of the question, at least for the moment.

Venus is completely different. It is almost as large as the Earth (if you represented Venus and the Earth by two snooker balls, a player could quite happily use them!), and it is very brilliant, partly because it is cloud-covered and partly because it is closer than any other natural body in the sky apart from the Moon. It can approach us to within 40,000,000 km, and at its best it may cast a shadow.

Yet telescopically, Venus is a disappointment, because there is almost nothing to see. The planet is surrounded by a dense, cloudy atmosphere which never clears, and all that can be made out is the brilliant disk with its phase – crescent, half, gibbous or almost full;

Venus is actually brightest when a crescent. A few vague patches can sometimes be made out, but they are not permanent, and they are difficult to draw. Before the Space Age, Venus was often called 'the planet of mystery', and we knew nothing about the conditions on its surface. It was sometimes thought – again wrongly – that the rotation was captured or synchronous, equal to Venus' 'year' of 224.7 Earth-days.

Obviously it was hot. It is some 40,000,000 km closer to the Sun than we are, and in addition the atmosphere was known to contain a great deal of the heavy gas carbon dioxide, which acts in the manner of a greenhouse and tends to shut in the Sun's heat. But we knew little else; Venus might be a scorching dust-desert, or it might be covered with water.

The first successful space-craft to Venus was Mariner 2, which by-passed the planet in December 1962 and sent back valuable data. In particular, it found that the surface is far too hot for water to exist, and that there is no measurable magnetic field. Subsequently, Soviet probes made controlled landings, and were able to send back images from the actual surface, and in 1989 the Americans launched the Magellan probe, which orbited Venus for

114. The surface of Venus, from the Russian probe Venera 9 which landed there on 21 October 1975 and transmitted for 53 minutes before being put permanently out of action by the intensely hostile conditions. Part of the space-craft is shown in the desolate rock-strewn landscape.

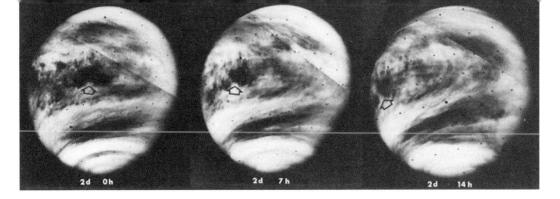

2d 0h   2d 7h   2d 14h

115. The rotation of Venus. These pictures were taken from the Pioneer space-craft in 1978. The arrows show the same regions, confirming the 4-day retrograde rotation period of the upper clouds.

years and drew up a radar map of the entire surface.

Almost everything found out was unexpected. The rotation period is 243 Earth days, longer than the orbital period, and Venus spins in a retrograde direction, so that if the Sun could be seen from the surface it would rise in the west and set in the east. The mean surface temperature is +480°C, but the upper clouds are very cold; there is little difference between the sunlit and dark hemispheres. The upper clouds spin much more quickly than the surface, with a rotation period of only 4 days. The atmosphere is indeed made up chiefly of carbon dioxide, but the clouds contain large amounts of sulphuric acid. It has been said that conditions on Venus are very like the usual idea of hell.

The radar map due to the Magellan orbiter shows that Venus is a world of plains, highlands and lowlands, with lowlands accounting for 65 per cent of the surface; there are two main highland areas, known as Ishtar and Aphrodite; Ishtar is about the size of Australia. The highest mountains (the Maxwell Mountains), adjoining Ishtar, rise to 11 km above the mean radius of the planet. Volcanic activity has left its mark all over Venus; there are structures ranging from small domes up to huge shield volcanoes, of the same type as Earth's volcanoes in Hawaii, but much larger. Lava-flows are everywhere, and it is very likely that volcanic eruptions are going on at the present time. It seems that life there is out

of the question, and the chances of any manned expeditions in the foreseeable future seem to be nil.

Why is Venus so like the Earth in size and mass, and so unlike it in every other way? The answer must lie in Venus' lesser distance from the Sun. When the two worlds were formed, around 4.6 thousand million years ago, the Sun was not so luminous as it is now, and Venus and Earth may well have started to evolve along similar lines. Then, however, the Sun became more powerful. Earth was moving at a safe distance, but Venus was not. The oceans boiled away; there was a runaway greenhouse effect, and the carbonates were driven out of the rocks to produce the present choking atmosphere. Any life which may have gained a foothold was destroyed. It is rather sobering to reflect that if Earth

116. The surface of Venus, seen from the Magellan probe. Lava-flows are much in evidence.

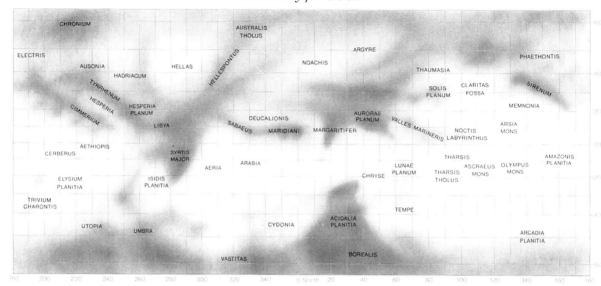

117. Map of Mars drawn from observations by Patrick Moore. This is on a Mercator projection. South is to the top. The main telescope used was a 39-cm reflector.

had been much closer to the Sun, the same thing would have happened here – and you would not now be reading this book.

Beyond the Earth-Moon system we come to Mars, which can approach us to within 60,000,000 km – not as close as Venus, of course, but Mars is much easier to observe, because when at opposition we can see the whole of its disk, and the phase is always greater than 85 per cent. In size and mass it is intermediate between the Earth and the Moon, and we would therefore expect it to have a thin atmosphere – which is precisely what we find. It is a colder world than the Earth, but it is not permanently frozen, and before the Space Age it was thought that life, in some form, might exist there.

Because Mars is relatively small, it can be studied in detail for only a few months to either side of opposition, and a fairly powerful telescope is needed to show much upon its red disk. However, there are obvious dark patches, which, unlike the cloudy features of Venus, are permanent,

and can be mapped. The most conspicuous of them is the V-shaped Syrtis Major, which is well within the range of a 3-inch (7.6-cm) refractor under good conditions. The poles are covered with white caps, which change with the Martian seasons – largest in winter, smallest in summer. Mars takes 687 Earth-days to go once round the Sun, but this is equivalent to only 668 Martian days or 'sols', because Mars spins more slowly than the Earth; a

118. The Sojourner 'rover' at Ares Vallis, on Mars.

119. Mariner 9, atop an Atlas-Centaur rocket, is launched toward Mars on 30 May 1971. It entered Martian orbit on 13 November, and operated until 27 October 1972, by which time it had returned 7,329 excellent pictures.

were not sea-beds; they were merely regions where the reddish 'dust' had been swept away by Martian winds, exposing the darker rocks underneath. The atmosphere was much thinner than had been expected, with a density much less than that of the Earth's air at the top of Everest (we now know that the ground pressure is below 10 millibars everywhere). Moreover, the atmosphere is made up mainly of carbon dioxide. There were craters, looking rather like those of the Moon, and there was absolutely no sign of life. Certainly there was no trace of the famous canals, straight features crossing the deserts, which had been reported by many observers and which had often been regarded as artificial waterways.

Mariner 4 was only the first of many probes to Mars. In 1971 Mariner 9 was put into orbit round the planet, and sent back magnificent pictures of the craters, mountains and valleys; the highest volcano, aptly named Olympus Mons (Mount Olympus) is 25 kilometres high, with a huge central crater. There are other volcanoes, almost equally lofty, and there

120. Craters on Mars, as shown by Mariner 9 in 1971. This was the first space-craft to send back detailed views of the volcanic areas.

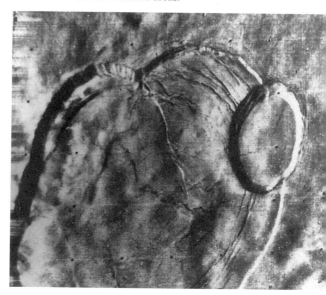

day there is equal to 24 hours 37½ minutes. The tilt of the axis is almost the same as ours, so that the seasons are of the same type, though they are of course much longer.

The map given opposite (Fig. 117) was drawn from my observations made with 39-cm and 31-cm reflectors. The main dark and bright areas are well-defined; until the 1960s it was thought that the dark regions were old sea-beds filled with low-type vegetation, while the red areas were dusty deserts. It was also thought that the surface was fairly smooth, with no high mountains or deep valleys.

How wrong we were! In 1965 the American space-craft Mariner 4 passed by the planet at a distance of only 10,000 km, and sent back a mass of information, not all of which was welcome. The dark areas

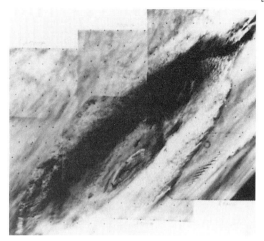

121. Oblique view of the great volcano Olympus Mons, on Mars; picture from Viking 1.

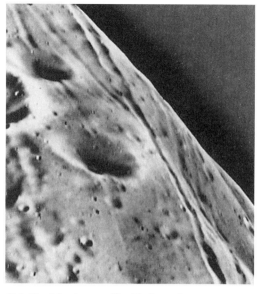

123. The Martian satellite Phobos, photographed by Viking Orbiter 1 on 20 February 1977 at a range of 120 km. The picture shows a region in the northern hemisphere that has striations and is heavily cratered.

are deep basins as well, one of which (Hellas) measures 2200 x 1800 km. The longest system of valleys is known as the Valles Marineris (Mariner Valley), over 4500 km long and 7 km deep.

Then, in 1976, two American probes – the Vikings – made controlled landings on Mars, and were able to send back direct pictures of the red, rock-strewn landscape. In the Martian day, the sky was pinkish, and the temperature was very low, reaching a maximum of no more than –31°C. There are clouds in the atmosphere,

and there are occasional dust-storms which spread over the entire planet.

The Vikings and Mariners showed features which are almost certainly dry riverbeds, in which case conditions there must have been much less unfriendly in the remote past. The polar caps are made up of a mixture of water ice and 'dry ice' – solid carbon dioxide.

The Viking landers scooped up material from the Martian surface, analysed it, and sent back the results. Again no definite traces of life were found.

122. High-resolution photograph of the surface of Mars, taken by Viking Lander 2 at its Utopia Planitia landing site on 18 May 1979, and relayed to Earth by Orbiter 1 on 7 June. It shows a thin coating of water ice on the rocks and soil.

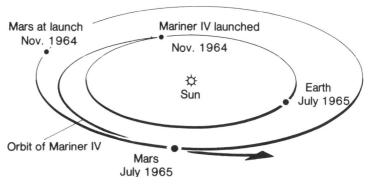

Mars at launch
Nov. 1964

Mariner IV launched
Nov. 1964

☼
Sun

Earth
July 1965

Orbit of Mariner IV

Mars
July 1965

124. Transfer orbit to Mars. This shows the path of the first successful Mars probe, Mariner IV.

Following the Viking success, there were various attempts to send further probes to Mars. Several of these failed. Then, on 4 July 1997 – America's Independence Day – the Pathfinder space-craft came down in the area of Mars known as Ares Vallis. This time there was no attempt at a 'soft' impact. Pathfinder was encased in airbags, and simply bounced around before coming to rest. It carried a small 'rover', Sojourner (Fig. 118), which was equipped with wheels; it crawled out of Pathfinder, and was able to move around, guided by controllers in NASA. Rocks were analysed, and established that the area really was an old flood-plain, so that in the past Mars did have oceans. Sojourner continued to transmit information until 27 September.

The Mars Global Surveyor probe (launched 1996) entered a closed orbit round Mars, and provided improved maps. However, Mars Climate Orbiter (launched December 1998) failed; the cause was human error. At NASA, one team used Imperial measure for a key space-craft operation, while another used Metric units. The result was that Climate Orbiter crashed on to the surface of the planet. Mars Polar Orbiter was designed to land on Mars in December 1999, and to search for traces of life. Unfortunately it too failed. Further missions are planned, but whether any traces will be found remains to be seen. It may be too early to dismiss Mars as being completely sterile.

It has been claimed that certain meteorites, found in Antarctica, were blasted away from Mars, and contain traces of very small and primitive organisms which could be classed as 'life'. However, the evidence is, at best, very inconclusive.

Mars has two satellites, Phobos and Deimos. Both are very small – Phobos less than 30 km in diameter, Deimos less than 20 km – and both are irregular, cratered and rocky, as the Viking and Mariner pictures showed. They are quite unlike our massive Moon, and may be asteroids which were captured by Mars long ago. Phobos takes only 7 hours 39 minutes to complete one orbit round Mars, and is very close to the surface – no further away than London is from Aden. To an observer on the planet, Phobos would rise in the west, cross the sky in only 4½ hours, and set in the east, while the interval between successive risings would be no more than eleven hours. In 1988 the Russians launched two space-craft to study Phobos from close range, but both failed: contact with them was lost, and was never regained.

Manned missions to Mars are being planned, and may well take place within the next twenty or thirty years. Since the atmosphere is so thin, it will never be possible to walk around in the open without wearing space-suits, but at least Mars is less unwelcoming than Mercury or Venus, and it must surely be the first world beyond the Moon which we can hope to reach.

The main problem, of course, is that the journey will take a comparatively long time. To wait until Mars is at its closest to us, and then simply fire a rocket across the gap, would mean using more fuel than could possibly be carried. What has to be done is to speed the rocket up, relative to the Earth, so that it swings outward, reaching the orbit of Mars at a pre-arranged point (Fig. 124); it will make most of the journey unpowered, in free fall, following what is termed a transfer orbit. Even so, the journey will take months, and there can be no question of a quick 'there-and-back' trip, as was possible with the Moon.

This means, too, that some sort of a base will have to be set up at once. What form it will take remains to be seen. Remember, too, that the surface gravity is only 0.38 that of the Earth, so that colonists who spend a long time there may have difficulties on returning to Earth and becoming 'heavy' again. These are problems for the future, but sooner or later they will have to be tackled.

Beyond the orbit of Mars there is a wide gap before we come to Jupiter, the first of the giant planets. Astronomers of the late eighteenth century suspected that there might be a planet there, mainly because of a relationship known as Bode's Law, which is associated with the distances of the planets from the Sun.* In 1801 G. Piazzi, from Sicily, discovered a small body in just about the right orbit. It was named Ceres, and was insignificant by planetary standards; its diameter is now known to be only 940 km, so that it was no real surprise when three more similar bodies were found within the next few years – Pallas, Juno and Vesta. Collectively, they became known as the asteroids or minor planets. A fifth

---

* Take the numbers 0, 3, 6, 12, 24, 48, and 96, each of which (after 3) is double its predecessor. Add 4 to each, giving 4, 7, 10, 16, 28, 52, and 100. If the Earth-Sun distance is taken as 10, the series gives the distances of the other planets, to scale, with fair accuracy: the distance of Mercury is 3.9, Venus 7.2, Mars 15.2, Jupiter 52.0 and Saturn 95.4. There was no planet corresponding to the Bode number 28, which was why astronomers of the time thought it worth while to begin a systematic search; Ceres has a 'Bode distance' of 27.7. However, though the Law works quite well for Uranus, it breaks down completely for Neptune and Pluto, and it is probably due to nothing more than sheer coincidence.

was found in 1845, and more discoveries soon followed, so that by now the total number of known asteroids has risen to over 13,000. Very few are more than 50 km in diameter, and only one (Vesta) is ever visible with the naked eye.

No asteroid is massive enough to hold on to any trace of atmosphere. It seems likely that no large planet could form in that part of the Solar System, because the powerful pull of Jupiter would break it up before it had been able to condense. The total membership of the swarm may be at least 100,000, and probably more.

Several asteroids have been imaged by space-probes, and, as expected, have proved to be cratered. Collisions between asteroids must be frequent, and only the largest members of the swarm are spherical in shape.

Some small asteroids move in eccentric orbits which carry them away from the main zone. Such is Eros, a sausage-shaped body less than 30 km long, which can approach the Earth to within 25,000,000 km. In 1999/2000 it was surveyed by the spacecraft NEAR (Shoemaker). In 1937 an even smaller asteroid, Hermes, brushed past us at only about twice the distance of the Moon, and since then many other 'close-approach' asteroids have been found. Two known asteroids, Icarus and Phaethon, have paths which carry them even closer to the Sun than the orbit of Mercury, so that at perihelion they must be red-hot. On the other hand, the so-called Trojan asteroids move in the same path as Jupiter, keeping either well ahead of or well behind the giant planet. Asteroid No. 944, Hidalgo, travels out almost as far as Saturn, while No. 2060, Chiron, spends most of its time between the orbits of Saturn and Uranus. The nature of Chiron is uncertain. In 1989 it was reported that a gas-and-dust cloud had developed round it, and it is possible that Chiron may be more cometary than asteroidal.) Others of the same type have also been found. The orbits of some of these exceptional asteroids are shown in Fig. 125.

Of course, there is always the chance

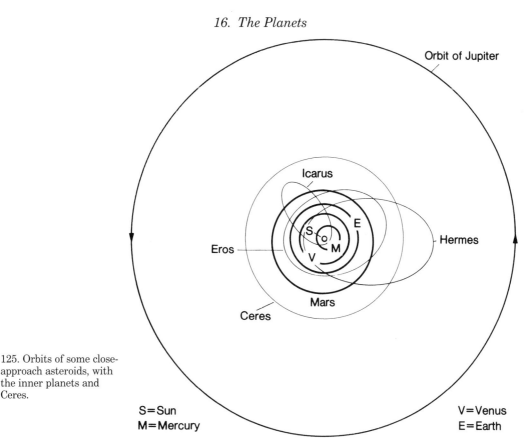

Orbit of Jupiter

Icarus

Hermes

Eros

Mars

Ceres

125. Orbits of some close-approach asteroids, with the inner planets and Ceres.

S = Sun
M = Mercury

V = Venus
E = Earth

that the Earth will be hit by a wandering asteroid, and there is a theory that this did happen some 65 million years ago, changing our climate so markedly that the dinosaurs could not cope with the new conditions, and died out. Whether this is true or not, there is no need for alarm. The danger of collision with an asteroid in the foreseeable future is very slight indeed.

The giant planets, far out beyond the main asteroid zone, are quite unlike the worlds we have talked about so far. They have gaseous surfaces, and in some respects we can consider them all together, though they are by no means identical – and in particular, the Jupiter/Saturn pair is very different from the Uranus/Neptune pair.

Much of what we know has been drawn from four space-craft, Pioneers 10 and 11 of the 1970s, and Voyagers 1 and 2. There is no point in saying much about the

Pioneers, because the Voyagers were much more effective. Their tracks through the Solar System are shown in Fig. 126.

Voyager 1 was launched on 5 September 1977, and swung out to pass Jupiter in March 1979. It then used Jupiter's powerful pull of gravity to put it into an orbit towards Saturn, which it by-passed in November 1980, after which it began a never-ending journey which will take it out of the Solar System for ever.

Voyager 2 was more ambitious. It started its journey a few days before its twin, but it was travelling in a less economical path, and did not reach the neighbourhood of Jupiter until July 1979. It then went on to Saturn, making its pass in August 1981. Saturn's gravitational pull was then used to send it on to Uranus in January 1986; in turn, Uranus sent it on to encounter with Neptune in August 1989. This 'gravity

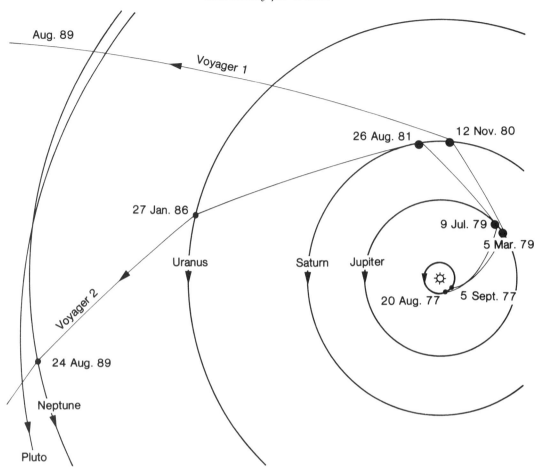

Aug. 89

Voyager 1

12 Nov. 80

26 Aug. 81

27 Jan. 86

9 Jul. 79

5 Mar. 79

Uranus

Saturn   Jupiter

Voyager 2

20 Aug. 77   5 Sept. 77

24 Aug. 89

Neptune

Pluto

126. Trajectories of Voyagers 1 and 2. Neptune was actually by-passed by Voyager 2 in August 1989, slightly ahead of the original schedule.

assist technique' was possible only because in the late 1970s the giant planets were arranged in a sort of curve. This will not happen again for well over a century.

The Galileo probe, launched in October 1989, reached Jupiter in 1995. It consisted of two sections: an orbiter, to enter a closed path round the planet and send back data about Jupiter itself and about its satellites, and an entry probe, which would plunge into the Jovian clouds and send back information until being destroyed. The entry took place on 7 December 1995, and was very successful, sending back valuable details about conditions below the visible surface. The orbiter was also successful, and was still sending back data at the end of 2000.

Jupiter is much the largest and most massive planet in the Solar System. According to the latest models, it has a rocky core at high temperature – at least 30,000°C – surrounded by layers of liquid hydrogen, above which comes the deep, cloudy atmosphere, made up chiefly of hydrogen and helium. Jupiter sends out more energy than it would do if it depended entirely upon what it receives from the Sun, but that is not to say that it is at all like a star; its core is not nearly hot enough to trigger off nuclear reactions. There is a powerful magnetic field, and the space-probes have also shown that the planet is surrounded by zones of radiation strong enough to kill any astronaut foolish enough to venture inside them. Jupiter is also a source of radio waves.

The overall colour is yellowish. Any casual look through a telescope will show that the disk is flattened; the diameter is approximately 144,000 km as measured through the equator, but only 134,000 km when measured through the poles. The reason is that Jupiter spins very quickly on its axis – the 'day' there is less than ten hours long – so that the equator bulges out. Neither does Jupiter rotate in the way that a solid body would do. The equatorial zone has a rotation period about five minutes shorter than that of the rest of the planet, and various markings on the disk have periods of their own, so that they drift around in longitude.

The most striking features are the dark belts, of which several may be seen with a 3-inch (7.6-cm) refractor. Generally there are two which are particularly prominent, one to the north of the equator and the other to the south (Fig. 127). There are also brighter zones. It seems that the bright zones are regions where gas is rising from below, while the dark belts are regions where the gas is descending. The clouds are not the same as ours, and are made up largely of droplets of ammonia.

Spots are frequent. Most of them are fairly short-lived, but there is one exception: the Great Red Spot, a huge oval which, at its maximum size, has a surface area greater than that of the Earth. It has been under observation for several centuries, and at times it really is red – coloured, probably, by phosphorus. At times it disappears for a while, but it always comes back, and its position can usually be traced by the hollow which it forms in the south equatorial belt. The space-probes have shown that it is a whirling storm – a phenomenon of Jovian 'weather'.

With an adequate telescope, Jupiter shows an amazing amount of detail. As well as the belts and spots, there are many more delicate features which shift and change over short periods. Moreover, Jupiter is spinning so quickly on its axis that any particular feature will be carried across the disk from one side to the other in only a few hours. To make the view even more interesting there are the four bright satellites, known collectively as the Galileans because the first systematic observations of them were made by Galileo in 1610. Of these, Io is slightly larger than our Moon, Europa slightly smaller, and Ganymede and Callisto much larger; indeed Ganymede, with its diameter of over 5000 km, is larger than the planet Mercury, though admittedly it is not so massive.

127. Belts and zones of Jupiter.

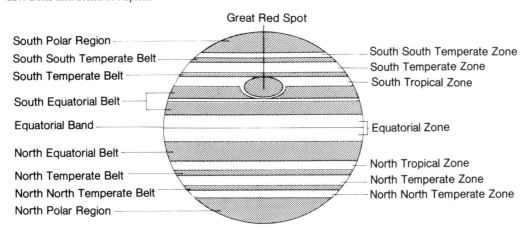

128. Io, imaged from the Galileo probe in June 1999. The volcanic structures are well shown.

130. A shadow transit of Ganymede across Jupiter's disk. The shadow of Ganymede is seen just below the Red Spot.

It is fascinating to watch the Galileans as they move round Jupiter, showing all sorts of phenomena (Fig. 130). They may be seen in transit against Jupiter's disk; there may be shadow transits; the satellites themselves may be eclipsed by Jupiter's shadow, or occulted behind the planet. A good project for the owner of a telescope is to track them from night to night, and see how they behave.

All four Galileans were studied from close range by the Voyagers and Galileo. Ganymede and Callisto are icy and cratered; Europa is icy and smooth; Io is red, with a sulphur-covered surface and violently active sulphur volcanoes; a strong electric current connects Io with Jupiter, and the orbital position of Io has a marked effect upon the Jovian radio emission. Europa has an ice-covered surface, and there are indications that below this there may be an ocean of liquid water – though the chances of finding any life under such conditions do not appear to be very high!

129. Surface of Europa: Galileo space-craft, May 1999. The complex nature of the icy crust is very evident. The prominent dark bands are up to 20 km wide, and may have been formed when the crust fractured, separated, and was filled in with darker, dirtier ice or 'slush' from below.

There are twelve other satellites, but all are too small and faint to be seen with small telescopes. The outer four move round Jupiter in a wrong-way or retrograde direction, so that they may be captured asteroids rather than true satellites. There is also a thin, dark ring.

Saturn, smaller and further away than Jupiter, has a globe of the same basic type. There are belts, bright zones and occasional

bright white spots, though in general the disk shows much less detail than that of Jupiter. Like Jupiter, it seems to have a solid core, overlaid by layers of liquid hydrogen and then by the gaseous atmosphere, but the overall density is less – and indeed, the mean density of Saturn is less than that of water. There is a magnetic field, and there are also radiation zones, but these are much less powerful than Jupiter's.

The main glory of Saturn lies in its ring system. There are three main rings, two of which are bright and the other semi-transparent (Fig. 132); the two bright rings are separated by a gap known as Cassini's Division in honour of the Italian astronomer who discovered it as long ago as

131. Jupiter, from Voyager 2; more detail is shown than could ever be made out by using Earth-based telescopes.

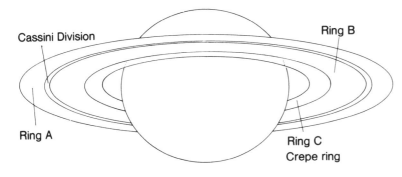

Cassini Division

Ring B

Ring A

Ring C
Crepe ring

132. The main ring system of Saturn; the bright rings A and B, separated by the Cassini Division, and the semi-transparent ring C, or the Crepe Ring.

1675. Though the rings look solid when seen through a telescope, it has long been known that any solid or liquid ring would be quickly broken up by Saturn's pull of gravity. The rings are made up of swarms of icy particles, ranging in size from pebbles to large blocks.

The diameter of the ring system is almost 275,000 km, but the thickness is

133. Saturn. This mosaic of images was taken by Voyager 1 on 30 October 1980 from a distance of 18 million km. The soft, velvety appearance of the low-contrast banded structure and increased reflection of blue light near the perimeter of the Saturn disk are due to scattering by a haze layer above the planet's cloud deck. Features larger than 350 km are visible. The projected width of the rings at the centre of the disk is 10,000 km, which provides a scale for estimating feature sizes on the image.

probably no more than 1 km. When the rings are edgewise-on to us, as happened in 1995 and will again in 2009, they almost vanish; when best displayed, they are magnificent even when seen through a small telescope. The changing aspects of the rings are shown in Fig. 135.

The Voyagers showed that the rings are more complicated in form than had been expected; there are in fact thousands of narrow ringlets and minor divisions. Several new, faint rings were also found outside the main system. We are still not sure of the reason for this strange, 'grooved' appearance, but presumably the gravitational effects of Saturn's satellites are mainly responsible for it.

There are eighteen known satellites. Of these, Titan is bright enough to be seen with a small telescope; a 3-inch refractor will also show Iapetus and Rhea, and perhaps Dione and Tethys; the others are fainter. All were surveyed by the Voyagers, and have proved to be very different from each other. The outermost, Phoebe (220 km in diameter) has retrograde motion, and may be an ex-asteroid; Iapetus (1436 km) has one side which is as bright as snow and the other which is blacker than a

134. Titan, as imaged by the Hubble Space Telescope. Bright and darker areas are seen, but the nature of the surface is still not known.

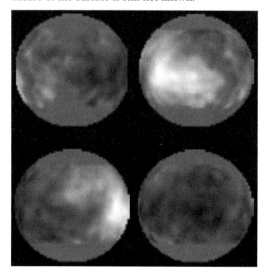

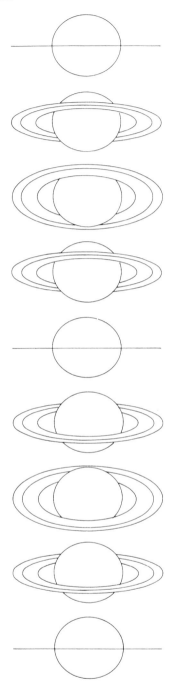

135. The changing aspect of Saturn's rings through a complete cycle.

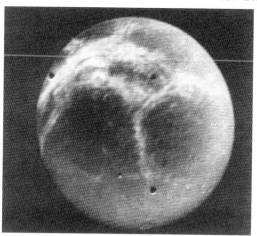

136. Saturn's icy moon Dione, taken by Voyager 1 from a range of 790,000 km on 12 November 1980. The bright wispy markings may be surface frost deposits. Circular impact craters up to 100 km in diameter can be seen

blackboard; Hyperion (300 km) is shaped like a hamburger, while Mimas (400 km) is dominated by a huge crater with one-third the diameter of the satellite itself.

The only really large satellite, Titan, is of special interest. It is over 5000 km across, and has a dense atmosphere, made up chiefly of nitrogen. Though the Voyagers could show nothing but the upper clouds, Titan was contacted in June 1989 by radar; the large radio telescope at Goldstone in the USA was used. Since then a certain amount of detail has been shown by the Hubble Space Telescope, and certainly Titan is unlike any other world in the Solar System. The atmosphere is made up chiefly of nitrogen, with a good deal of methane; the ground pressure is 1.3 times that of the Earth's air at sea level. There may well be pools or even seas of liquid ethane, but the chances of life are slim, if only because Titan is so cold; the surface temperature is around −180°C, which is near the triple point of methane (that is to say, methane could exist as a solid, liquid or gas, just as H_2O can do on Earth as ice, liquid water or water vapour).

At present a new probe, Cassini, is on its way to the Saturnian system, and will arrive there in 2004. It carries a smaller vehicle, Huygens, which is designed to make a controlled landing on Titan and send back data from the surface. Whether Huygens will come down on firm ground, or splash into a chemical ocean, remains to be seen.

Saturn was the outermost of the planets known in ancient times, and it came as a surprise when, in 1781, William Herschel discovered the planet we now call Uranus. He was not looking for a planet; he was carrying out a 'review of the heavens' with a home-made reflector, and even when he chanced upon Uranus he did not know what it was. However, he knew that it could not be a star, because it showed a definite disk and moved slowly from night to night. He mistook it for a comet, but before long it was found to be a new member of the Sun's family, moving far beyond the path of Saturn.

Uranus is just visible with the naked eye, but you have to know where to look for it. No telescope will show much upon its pale, greenish disk. It has about half the diameter of Saturn, and is of rather different composition, with less hydrogen but more ammonia and water.

One strange feature of Uranus is the tilt of its axis, which is as much as 98° to the orbital plane – more than a right angle (Fig. 137). There are times when one or other of the poles faces the Sun, and has a 'day' as

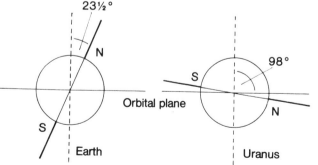

137. The strange axial tilt of Uranus – more than a right angle.

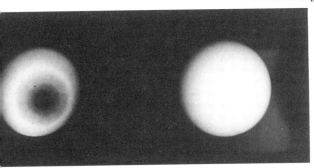

138. Uranus, from Voyager 2; January 1986. The left-hand image is processed so as to make comparisons of intensity easier.

long as 21 Earth-years. The reason for this is unknown.

In 1977 it was found that Uranus has a system of thin, dark, narrow rings. They are very difficult to detect from Earth, and were discovered more or less by accident. There are times when Uranus passes in front of a star, and hides or occults it. When this happened, the star was seen to wink regularly both before and after the actual occultation, as shown in Fig. 139, so that clearly it was being briefly blotted out by ring material. The rings of Uranus are not icy; indeed, they are as black as soot, and were not studied in detail until the Voyager 2 probe sent back excellent pictures of them in January 1986. Voyager also discovered a magnetic field, though the magnetic axis is inclined to the axis of rotation by 60°; there are radiation zones, and vague cloudy features were seen on the disk. Unlike the other giant planets, Uranus does not seem to have a strong source of internal heat.

Five satellites were known before the Voyager 2 mission: Miranda, Ariel, Umbriel,

139. Discovery of the rings of Uranus. On 10 March 1977 Uranus occulted a star, and the phenomenon was observed from the Kuiper Airborne Observatory – an aircraft carrying a powerful telescope. Before and after occultation the star 'winked', betraying the presence of ring material associated with Uranus.

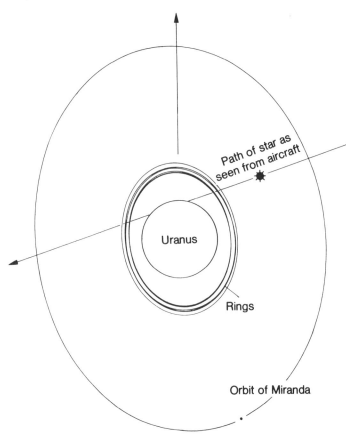

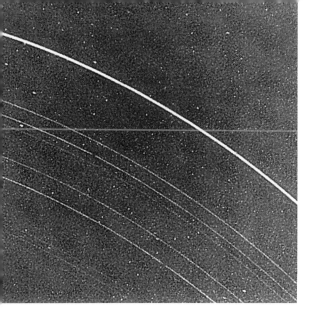

140. The rings of Uranus (Voyager 2, January 1986). They are quite unlike the brilliant, icy rings of Saturn.

Titania and Oberon. All are smaller than our Moon, and even the largest, Titania, is only about 1500 km in diameter. Voyager showed that they have icy, cratered

141. Miranda, the remarkable satellite of Uranus with its varied landscape; from Voyager 2, January 1986. (The notches to the top left are instrumental effects.)

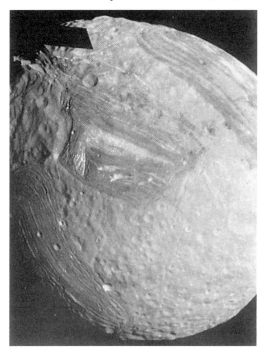

surfaces, but that of Miranda – only 470 km across – is amazingly varied, with ridges, ice cliffs, and large enclosures which have been nicknamed 'race-tracks'. Evidently it has had a violent history. Ten new satellites were discovered, but all are very small.

After Uranus had been under observation for some time after its discovery, mathematicians found that it was not moving as it had been expected to do. Something was pulling it out of position, and two men, John Couch Adams in England and Urbain Le Verrier in France, independently realised that this 'something' must be an unknown planet. In fact, Adams and Le Verrier were presented with a scientific detective problem; they could watch the behaviour of the victim (Uranus) and they had to track down the culprit. They worked out where the new planet should be, and then contacted national observatories to see whether it could be found. Adams' work was not followed up at once, but Le Verrier's calculations were sent to the Berlin Observatory in 1846, and almost immediately Johann Galle and Heinrich D'Arrest located the planet – almost exactly where Le Verrier had said it would be. It was named Neptune.

In size and mass Neptune is very similar to Uranus; it is only slightly smaller, but is appreciably more massive. However, in other ways the two outer giants are very different, as was confirmed when Voyager 2 made a close encounter with Neptune in August 1989; the space-craft passed over Neptune's north pole at a distance of only 5000 km. Neptune has a pronounced internal heat-source, and is much more active than the placid, almost featureless Uranus. Voyager showed a huge dark spot, now known generally as the Great Dark Spot, above which are high-altitude 'cirrus' clouds of methane; these clouds lie at around 50 km above the main cloud-deck. There are also other dark spots, and winds blowing in a retrograde direction at over 1000 km per hour. The overall rotation period of the planet is 16h 3m. Recently, images of Neptune have been obtained with

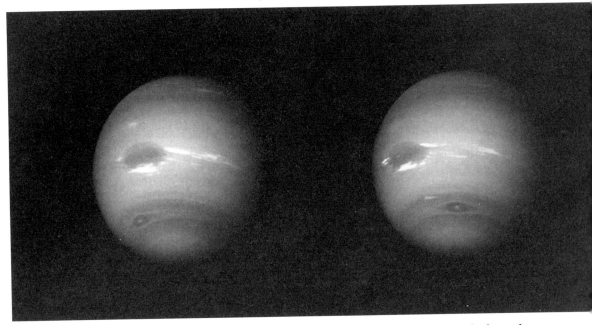

142. These two images of Neptune were taken from Voyager 2 from a range of 12,000,000 km, on 21 August 1989. During the 17.6 hours between the left and right images, the Great Dark Spot, at latitude 22°S, has shifted noticeably, as it has completed a little less than one complete rotation of Neptune. The smaller dark spot, at latitude 54°S, has completed a little more than one rotation; the relative velocity between the two spots is 100 metres per second. The diffuse white feature north of the GDS is near Neptune's equator.

the Hubble Space Telescope, and it seems that things have changed markedly since the Voyager 2 pass. The Great Dark Spot is no longer to be seen, but other features have appeared. Evidently Neptune's surface is much more variable than had been previously believed.

Neptune is a source of radio emissions. There is a magnetic field, rather weaker than those of the other giants, and – surprisingly – it was found that, as with Uranus, the magnetic axis is both offset and is strongly inclined to the axis of rotation. There are three rings, much too faint to be seen from Earth, one of which has several brighter segments or arcs; there is also one broad sheet of particles. The entire ring-system is very dark.

Two satellites were known before the Voyager mission. One (Triton) is large, and thought to have an appreciable atmosphere; it moves round Neptune in a retrograde direction. The other (Nereid) is small, and has a very eccentric orbit. Voyager 2 discovered six new satellites, all relatively close to the planet; the largest of them has a diameter of over 400 km, and is considerably larger than Nereid, though it is so near Neptune that it is unobservable from Earth.

Triton itself proved to be a remarkable world. It is 2720 km in diameter, smaller than our Moon, and has a surface which shows little surface relief (no more than a kilometre or so), with few craters but many irregular formations a few kilometres across. The atmosphere is very tenuous, and is made up mainly of nitrogen, with some methane at lower levels.

The surface is coated with a mixture of nitrogen ice and methane ice – and is the coldest place ever studied in the Solar System. It is thought likely that below the surface there is a layer of liquid nitrogen, and that when this penetrates upward to a region of sufficiently low pressure it

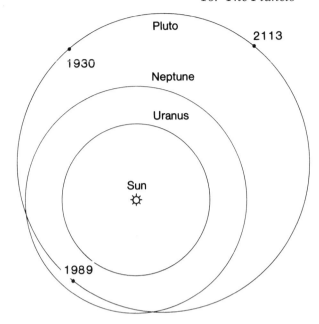

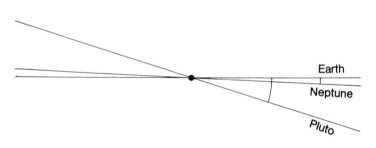

143. Orbit of Pluto. Though the mean distance from the Sun is much greater than Neptune's, the eccentric orbit brings it closer in near perihelion. Perihelion fell in 1989; it was not until 1999 that the distance of Pluto from the Sun again exceeded that of Neptune. There is no danger of collision with Neptune, as Pluto's orbit is inclined at 17°.

explodes, producing what may be called 'ice volcanoes' whose presence is shown by dark streaks coming from what are presumably vents. It also seems that Triton is not a true satellite of Neptune, but was once an independent body which was captured in the remote past.

Even when Neptune had been found, all was not well with the movements of the outer planets. Slight irregularities remained, and the problem was taken up by the American astronomer Percival Lowell, who had founded a major observatory at Flagstaff in Arizona (it was Lowell who had championed the theory that the canals of Mars were artificial waterways).

He was an expert mathematician, and he worked out a position for a new planet, but he failed to find it. In 1930, fourteen years after Lowell's death, Clyde Tombaugh, at Flagstaff, succeeded; after a photographic search he located the faint planet Pluto not far from the position which Lowell had given.

Pluto is an oddity. It is not a giant; it appears to be made up of a mixture of rock and ice, with only a thin atmosphere, and it is smaller and less massive than the Moon. Moreover, it has a peculiar orbit, much more eccentric than those of the other planets. Though its mean distance from the Sun is much greater than Neptune's, and it

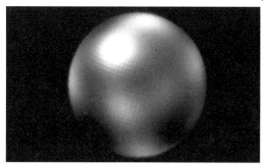

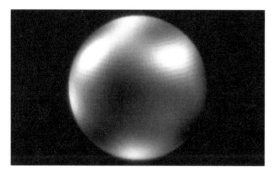

144. The surface of Pluto from the Hubble Space Telescope, showing light and darker areas.

has a revolution period of almost 248 years as against only 164.8 for Neptune, it spends part of its time closer-in to the Sun than Neptune can ever be (Fig. 143), though the orbit is also tilted at an angle of 17°, and there is no danger of collision. Pluto came to perihelion in 1989; between 1979 and 1999 Neptune, not Pluto, was the 'outermost planet'.

Yet is Pluto a true planet at all? In size it is more like a satellite, and it is not a solitary traveller, since it is accompanied by a secondary body, Charon, with a diameter nearly half that of Pluto itself (1199 km, as against 2445 km for Pluto). The distance between the two is no more than 19,000 km, and the revolution period of Charon is the same as the axial rotation period of Pluto: 6.3 Earth-days. To an observer on Pluto, Charon would seem to remain in a fixed position in the sky.

Because Pluto is so small and so lightweight, it could not possibly pull a giant planet such as Uranus or Neptune out of position to any measurable extent – and yet it was because of these effects that Lowell tracked it down. Either his reasonably correct prediction was sheer luck, or else the real planet for which he was hunting remains to be discovered.

Many asteroid-sized bodies have now been found, orbiting the Sun at distances about the same as that of Neptune, or beyond. These make up what has become known as the Kuiper Belt (after the Dutch astronomer G.P. Kuiper, who first suggested its existence). It has been claimed that Pluto is itself simply a large Kuiper Belt object rather than a true planet, and this may well be the case. Whether Lowell's 'Planet Ten' really exists, at a much greater distance, remains to be seen.

This has been a very sketchy tour of the Solar System, but at least I hope I have said enough to show you that each planet has a character and interest all its own. Use a telescope to look at the phases of Venus, the polar caps of Mars, the belts and moons of Jupiter and the rings of Saturn, and you will soon see what I mean.

16. The Planets

Questions

1. (a) What is so unusual about the axial rotation of Venus?
   (b) Why is it unlikely that Venus can support any form of life?
   (c) What did Mariner 10 tell us about the surface of Mercury?
2. (a) Is the atmosphere of Mars as dense as that of the Earth, and what is its composition?
   (b) How did the Viking probes carry out their search for life on Mars?
   (c) Why is it thought that there must once have been running water on Mars, even though there is none today?
3. (a) Describe the appearance of Jupiter as seen through a telescope.
   (b) What is Jupiter's Great Red Spot, and where on the planet's disk would you find it?
   (c) In what important way does Io differ from the other Galilean satellites of Jupiter?
4. (a) Why did some nineteenth-century astronomers believe that there should be a planet moving round the Sun between the orbits of Mars and Jupiter?
   (b) Voyager 2 by-passed all four giant planets by using the gravity-assist technique. Explain what this means.
   (c) How was the position of Neptune worked out before the planet was actually identified?
5. (a) Give a brief description of Saturn's ring system.
   (b) What is unusual about the axial rotation of Uranus?
   (c) Should Pluto be regarded as a true planet? If not, why not?

Practical work

Most practical projects concerning the planets mean using a telescope. If you have one, then make regular drawings of Venus, showing the changing phase, and also draw the dark markings and the polar caps of Mars (unless you have a fairly powerful telescope, say, 6-inch aperture or more, this will be possible only when Mars is fairly near opposition). Jupiter is more rewarding; note the changing aspects of the disk, and also record the positions of the Galilean satellites on different nights, noting their transits, shadow transits and eclipses. Make sketches of Saturn, and see whether you can make out the Cassini Division; you should be able to see Titan, and also some of the other satellites provided that you have a telescope of at least 3-inch (7.5-cm) aperture.

Uranus and Neptune can be located, but you will need to know their positions; look them up in a publication such as the *Yearbook of Astronomy*. Mercury can be seen with the naked eye when well placed; details about its visibility each month are given in astronomical periodicals. You may also be able to identify some of the brighter asteroids, though even with a telescope the asteroids look exactly like stars.

17
Comets, Meteors and Meteorites

Next, in our survey of the Sun's family, we come to its junior members. It is fitting to begin with the comets, which look much more important than they really are, and which used to cause a great deal of alarm in ancient times – partly because they were regarded as signs that the gods were angry with mankind, and partly because it was thought that a direct collision with a comet might mean the end of the world.

In fact, no comet could destroy the Earth, and even a direct hit would be inadequate to jolt the Earth out of its path round the Sun. This is because a comet is of very low mass, with an extremely weak pull of gravity. It is made up of a small, solid centre or nucleus, composed mainly of ice, together with thin gas and what we may call 'dust'. Since it lies far away from the Earth, it does not move quickly across the sky, and has to be watched for many hours before its shift can be detected at all. If you see something crawling obviously against the starry background, then it most certainly cannot be a comet. Comets shine by reflected sunlight, though it is true that when near perihelion the material is affected by solar radiation and sends out a certain amount of light on its own account.

Though comets are genuine members of the Solar System, their orbits are not like those of the planets, and most of them move in very eccentric paths, though there are a few exceptions. The inclinations of the orbits may also be high, and some comets have retrograde motion. In fact, comets are the most erratic members of the Solar System,

and their low masses mean that they are easily 'pulled around' by the gravitational action of the planets. Generally speaking, they are visible to us only when they are moving in the inner part of the system, and there are not many comets which have been followed out as far as the orbit of Saturn. Remember, too, that according to Kepler's Laws, a comet must move fastest when near perihelion, so that it spends most of its time creeping along at a great distance from the Sun.

A great comet is really spectacular. It is made up of a central *nucleus*, composed chiefly if ice, with a diameter of no more than a few kilometres. Surrounding the nucleus is the head or *coma*, made of dust and gas (see Fig. 145). Extending away from the coma there may be one or more tails, sometimes stretching right across the sky. Tails are of two types: gas ('ion') tails, which are more or less straight, and dust tails, which are curved. However, tails are not always to be seen, and many small comets never produce them, so that they look like nothing more than fuzzy patches of luminous cotton-wool in the sky.

Comets seem to be very ancient. According to a theory proposed by the Dutch astronomer Jan Oort, there is a whole 'cloud' of them moving round the Sun at a distance of at least a light-year – much too far away to be seen from Earth. If one member of the cloud is disturbed for any reason, it may break away and start to fall in toward the Sun. After a journey lasting for thousands or even millions of years, it will reach the inner, warmer part of the Solar System, and will start to

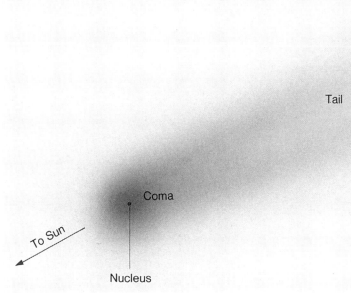

145. Structure of a comet, showing coma, nucleus and tail. Almost all comets move in very eccentric orbits.

become active. The ice in its nucleus will start to evaporate, forming the coma, and then the action of the Sun will drive thin material outward, producing a tail or tails. By the time it has reached the orbit of the Earth, it may have become really striking.

One of three things may then happen to it. It may fall into the Sun and be destroyed; by now photographs taken from artificial satellites have shown us several of these cometary suicides. It may

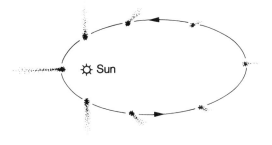

146. Direction of a comet's tail, which always points more or less away from the Sun.

be 'captured' by the action of a planet, usually Jupiter, and be forced into a smaller orbit in which it takes only a few years to complete one journey round the Sun. It may collide with a planet – and this did happen in July 1994. A comet, known as Shoemaker-Levy 9, actually impacted Jupiter, and caused disturbances in the Jovian gas which remained visible for months.

Or it may simply pass through perihelion and then begin its long journey back to the Cloud; as it recedes, and becomes cooler, the tails and coma disappear, and the comet reverts to its old state of a mere lump of ice, not to come back to perihelion for an immensely long period. This is why 'great' comets cannot be predicted, and are always apt to take us by surprise.

Though it seems fairly definite that long-period comets come from the Oort Cloud, it seems that the short-period comets, which take only a few years to go round the Sun, come from the much closer Kuiper Belt. We have said that a comet's tail is driven out of the coma by the action of the Sun. Light produces a pressure, which is very slight but is enough to drive gas-particles away from the coma; the

137

147. Bester's Comet of 1948 (1948 I), photographed by E.L. Johnson at the Union Observatory, Johannesburg. The comet was a naked-eye object, and was followed until February 1949. The orbit seems to be hyperbolic, in which case the comet will never return to the Sun.

solar wind, which we discussed earlier, drives out dust. It follows that the tails of comets always point more or less away from the Sun (Fig. 146), and when a comet is travelling outward, after having passed perihelion, it actually moves tail-first.

Great comets were quite common in the last century – in 1811, 1843, 1858, 1861 and 1882, for example – and some of them were brilliant enough to cast shadows; the coma of the Great Comet of 1811 was larger than the Sun, and the maximum length of the tail was 160,000,000 km, longer than the distance between the Earth and the Sun. Even brighter was the Great Comet of 1843, which was visible with the naked eye in broad daylight, and had a tail longer than the distance between the Sun and Mars. Of course, both these comets have periods of so many centuries that we have no hope of seeing them again, and their periods are very uncertain. No doubt they will return eventually, though it is true that on rare occasions a comet may be affected by the pull of a planet in a way that causes it to be thrown out of the Solar System altogether.

The twentieth century was relatively poor in great comets, but two appeared in its closing years. Comet Hyakutake became brilliant for a week or so during 1996; it will not return for 15,000 years. Then, in 1997, came Comet Hale-Bopp, which proved to be a magnificent spectacle, and which remained visible with the naked eye for months; it had a brilliant head and long tails. It was in fact a very large comet, with a nucleus perhaps 40 km in diameter, but unfortunately it did not come close to the Earth; its minimum distance was over 200 million km. Had it come as close as Hyakutake had done in the previous year, it would have cast shadows. It will not return to perihelion for more than 2000 years.

The short-period comets, which have been 'caught' from the Kuiper belt, have periods ranging from only 3.3 years (Encke's Comet) out to more than 150 years. They are seen regularly, and we always know when to expect them; Encke's Comet, first seen in 1786, has now been followed at over fifty returns to perihelion, and since it never moves as far out as Jupiter it can now be followed

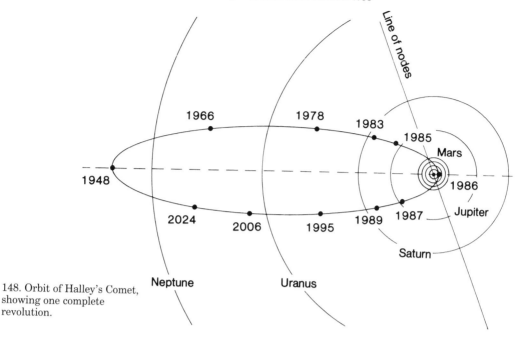

148. Orbit of Halley's Comet, showing one complete revolution.

throughout its orbit with large telescopes. However, most of the periodical comets are very faint. Each time a comet passes through perihelion it loses a portion of its mass, and gradually it becomes exhausted, so that periodical comets are short-lived by cosmical standards. Not many of them ever become visible with the naked eye, and many of them are lacking in tails of either type. It has even been suggested that some of the small close-approach asteroids, such as Icarus and Phaethon, may simply be ex-comets which have lost all their dust and gas.

A comet is usually named after its discoverer – for example, West's Comet of 1976 was discovered by the Danish astronomer Richard West – but there are others, such as Encke's Comet, which have been named in honour of the mathematician who first worked out the orbit. And this brings me to the most famous of all comets, Halley's.

In 1682 Edmond Halley, later to become Astronomer Royal, observed a bright comet, and measured its movement. At that time it

was generally thought (even by Newton) that comets came from outer space, passed by the Sun only once, and then vanished. Halley was not so sure. Looking back over the records, he found that the comet of 1682 moved in almost the same way as comets previously seen in 1607 and in 1531. Could they be one and the same comet, which came back every 76 years or so? This was what Halley suggested, in which case the comet would be seen once more in 1758. By then Halley had died, but the comet was recovered on Christmas Night 1758 by the German amateur astronomer Palitzsch, and since then it has returned in 1835, 1910 and 1986. Old reports have allowed us to trace it back as far as the return of 240 BC. It was also the comet seen in 1066, some months before the Battle of Hastings, and seems to have caused a great deal of alarm in King Harold's court; it is shown in the famous Bayeux Tapestry, said by some scholars to have been made on the orders of Matilda, William the Conqueror's wife.

The orbit of Halley's Comet is shown in Fig. 148. It is now back in the outer part of

149. The Giotto probe, half-built; Patrick Moore took this photograph of it at Bristol, on 19 January 1984, where the space-craft was being constructed. He is wearing a 'clean suit', as the whole area had to be kept completely dust-free.

the Solar System, and will next return to perihelion in 2061. Note that the period is not absolutely constant at 76 years; because of the pulls of the planets, no two orbits are exactly the same, and the period between one perihelion passage and the next may be as little as 74½ years or as great as 79 years.

At the last return, still fresh in the memories of many people, conditions were as bad as they could possibly be; when the comet should have been at its brightest, it was on the far side of the Sun – and although it was easily visible with the naked eye, it was not brilliant, as it had been in 1910. On the credit side, it was possible to send space-probes to it, and five were sent up: two Japanese, two Russian and one European. The European probe, named Giotto in honour of the Italian artist who had painted a famous picture of the comet at its return in 1301, passed right into the comet's head, and sent back

close-range photographs of the icy nucleus. The nucleus itself was small, with a longest diameter of 15 km, and – surprisingly – was covered with dark material; there were craters, hills, and jets of dust spurting out from below the surface on the comet's sunward side. Analysis showed that water ice made up at least 84 per cent of the nucleus.

It was calculated that at each perihelion passage, Halley's Comet must lose about 300,000,000 tons of material, and obviously it cannot survive for ever, though it will make many returns yet before its gases are exhausted. On the other hand, several small periodical comets which used to be seen regularly have now disappeared completely. Biela's Comet, with a period of 6.7 years, was seen at several returns, but in 1845 it split in two; the twins came back for the last time in 1852, but have never been seen since. As comets, they no longer exist, but in 1872 their débris was seen in the form of a spectacular shower of meteors, and this gave the final proof of the close connection between meteors and comets.

To recapitulate: a meteor is a tiny particle moving round the Sun. When it dashes into the Earth's upper air, it rubs against the air-particles, becomes heated by friction and burns away to produce the trail we call a shooting-star. What we see is not the meteor itself, but the luminous effects which it produces during its headlong flight toward the ground. It may be moving at anything up to 72 km per second. It cannot survive the full drop, but burns out well above ground level, ending its journey in the form of very fine 'dust'.

Meteors tend to travel in swarms, so that every time the Earth passes through a swarm we see a shower of shooting-stars. This happens many times in every year. The meteors in any particular shower are named from the constellation marking the *radiant* – that is to say, the position in the sky from which the meteors seem to come; thus the August shower, radiating from Perseus, is the Perseid shower, while the October Orionids come from Orion, and so on.

To show how this happens, consider the view from a bridge overlooking a motorway (Fig. 150). The parallel lanes of the motorway will seem to meet a point near the horizon, and cars coming down the lanes will seem to diverge from this distant point. (This is assuming, of course, that all the cars are travelling in the same direction

150. Overlooking a motorway. The parallel lanes seem to 'radiate' from a point near the horizon.

– something we would not recommend you to try!) It is much the same with the meteors in a shower, which are moving through space in parallel paths. Each shower has its own radiant. If you observe meteors at different times, and then plot their paths (Fig. 151), you will be able to identify the radiant.

As a comet moves, it leaves a 'dusty trail', and this is responsible for the meteor shower associated with it. For example, the October Orionids are the débris of Halley's Comet, while the Perseids are associated with Comet Swift-Tuttle, which has a period of 130 years, and last returned to perihelion in 1992.

The August Perseids are very reliable. Anyone who stares up into a dark, cloudless sky between 27 July and 17 August will be unlucky not to see at least a couple of meteors over a period of a few minutes. Other showers are less consistent. The Leonids of 17 November can produce superb displays, as they did in 1966, but in other years they are very few and far between. The 1999 display was poorly observed from England due to cloud. Obviously, then, the meteors of the Leonid swarm are bunched up, so that we have to hit the main swarm at just the right time, while the Perseids are spread out along the whole of the orbit of their parent comet (Fig. 152). Quite apart from shower meteors, we also have *sporadic* meteors,

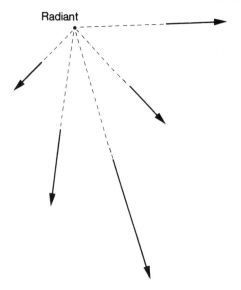

Radiant

151. Meteors of a shower diverge from a radiant, because their paths through space are in fact parallel.

153. The Hoba West Meteorite in Southern Africa, the largest known; it weighs over 60 tons. Photograph by Ludolf Meyer.

which are not associated with known comets, and may appear from any direction at any moment.

Now and then the Earth is hit by a larger object, which survives the complete drop to the ground without being destroyed, and is then called a *meteorite*. Most museums have meteorite collections. The meteorites are of various sizes, ranging from tiny particles up to huge blocks weighing tons; some are mainly stony in composition, while others are rich in iron.

It is quite wrong to assume that a meteorite is simply a large meteor. There is a basic difference; meteorites are not connected with comets, and they seem to be more nearly related to the asteroids –

in which case there is probably no difference between a large meteorite and a small asteroid.

Major falls are rare, and there is no reliable record of anyone having been killed by a plunging meteorite, though one or two people have had narrow escapes. Very occasionally a meteorite may produce a crater; of these the most famous is the Meteor Crater in Arizona (it really should be called the Meteor*ite* Crater), which is nearly 1300 m in diameter and 175 m deep. It was formed over 50,000 years ago, and is a well-known tourist attraction (if you find yourself in that part of the world, we recommend you to go and see it; it is an

152. With an old meteor stream (left) the cometary débris is spread all round the orbit. With a younger stream (right) this has not happened, and we see a major display only when we go through the densest part of the débris.

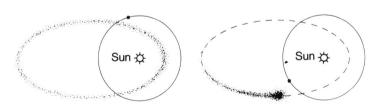

142

easy drive from Highway 66). In 1908 a missile struck the Tunguska region of Siberia and blew pine-trees flat over a wide area, but no crater was formed, and it may be that the object was mainly composed of ice, in which case it was more probably a fragment of a small comet. If it had landed in a populated area, it would have caused heavy damage and casualties.

Finally, we know that there are tremendous quantities of thinly-spread material in the Solar System. There are *micromete-orites*, which are too small to cause shooting-star effects when they dash into the upper air; there is material lying in the main plane of the Solar System, which can be lit up by the Sun to produce the cone-shaped glow called the *Zodiacal Light*, sometimes visible after sunset or before dawn, as well as the much fainter glow opposite to the Sun which is generally known as the *Gegenschein*. All in all, there is plenty to interest us in the Sun's family.

Questions

1. (a) If you see an object moving quickly across the sky, can it be a comet? If not, why not?
   (b) Why does a comet's tail point more or less away from the Sun?
   (c) Name the bright comet of 1997.

2. (a) How was Edmond Halley able to show that the comet of 1682 was periodical?
   (b) Is the period of Halley's Comet exactly 76 years? If not, why not?
   (c) Generally speaking, why are comets such as those of 1811 and 1843 brighter than periodical comets such as Encke's?

3. (a) Describe the main features of a great comet (use diagrams).
   (b) How are comets related to meteor streams?
   (c) In what ways do the orbits of most comets differ so strongly from those of planets?

4. (a) Explain, with a diagram, why the meteors of any particular shower seem to issue from a well-marked radiant point.
   (b) What are sporadic meteors?
   (c) What is the difference between a meteor and a meteorite?

Practical work

So far as comets are concerned, one has to wait for Nature to provide a suitable comet to observe. With binoculars, fainter comets can be identified if their positions are known, but it is necessary to consult some publication such as the Handbook of the British Astronomical Association. If you locate a comet, draw it as accurately as possible, and note its position against the background of stars.

Meteors are more accessible, and it is a fact that naked-eye observations of meteor showers are scientifically valuable. Some of the main annual showers are given in the table.

Name	Duration	Maximum	ZHR	Parent comet
Quadrantids	1-6 Jan	3 Jan	110	–
Eta Aquarids	2-7 May	4 May	20	Halley
Perseids	27 Jly-17 Aug	12 Aug	68	Swift-Tuttle
Orionids	16-26 Oct	21 Oct	30	Halley
Leonids	15-19 Nov	17 Nov	var.	Tempel-Tuttle
Geminids	7-16 Dec	14 Dec	58	–

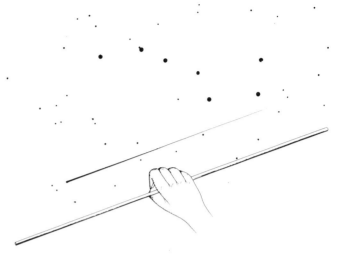

154. Plotting a meteor trail: hold up a stick against the position in the sky where the meteor had appeared.

The Quadrantids have no known parent comet; the radiant lies in the constellation of Boötes (the Herdsman), once separated out as the now-rejected constellation of Quadrans (the Quadrant). The Geminids also lack a parent comet, though it may be significant that they move in much the same orbit as the close-approach asteroid Phaethon. Usually the Leonids are sparse, but occasionally they may give magnificent displays.

The ZHR (Zenithal Hourly Rate) is the number of shower meteors which would be expected to be seen by a naked-eye observer under ideal conditions, with the radiant at the zenith or overhead point. In practice, these conditions are never fulfilled, so that the

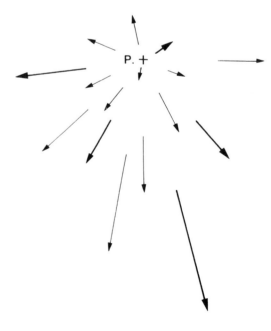

155. The meteors can then be traced back to point P, which is the shower radiant.

actual observed rate is always less than the ZHR.

The main task is to track the paths of the meteors, and determine their radiant. To do this, hold a stick up against the starry background along the track where the meteor has been seen (Fig. 154), and plot it on your chart. Note the time when the meteor was first seen, the track, any trail, and special features such as magnitude and 'flaring'. Obviously, a good knowledge of the constellation patterns is a great help. When you have finished your observations and have plotted them, the radiant of the shower should be fairly obvious (Fig. 155), but there will be some tracks which do not fit in; these will be due to sporadic meteors.

18
The Constellations

Several times during this book we have said that one important first step is to go outdoors on a clear night and learn your way around the sky. Now is the time to do it. We can do no more here than point out some of the most obvious of the constellations, but there are plenty of maps and books which will give you more detail – and believe us, it does not take very long. To all intents and purposes, the constellation patterns are permanent. Edmond Halley established that there are very slight shifts or *proper* motions detectable with the naked eye over periods of many centuries, but for the moment we can safely forget about them.

A star's apparent brightness is given by what is termed *apparent magnitude* (there are other kinds of magnitudes, but we will come to those later). The lower the magnitude, the brighter the star; as in golf, the more brilliant performers have the lower values, so that, say, a star of magnitude 1 is brighter than a star of magnitude 2. The naked eye can show stars down to about magnitude 6; a 39-cm reflector will take you down to about +15, and the faintest stars detectable by electronic devices on giant telescopes are of around +26. At the other end of the scale, we have magnitudes brighter than 1; the most brilliant star in the sky, Sirius, is –1.5 (to be more accurate, –1.46) and Venus can reach –4.4, while the apparent magnitude of the Sun is –26.8.

Conventionally, the 21 brightest stars are said to be of the 'first magnitude', ranging from Sirius (–1.46) to Regulus in Leo, the Lion (1.35). Since the naked eye can hardly distinguish differences of less than 0.1 magnitude, we propose to give the values here to the nearest tenth, so that Sirius will be taken as –1.5 and Regulus as 1.3. (Unless otherwise noted, all magnitudes are assumed to be positive – that is to say, fainter than zero. Only four stars are brighter than this: Sirius, Canopus (–0.7), Alpha Centauri (–0.3) and Arcturus (–0.04).)

Since we are writing for GCSE students in Britain, we will deal only with the constellations visible from our islands, which unfortunately means cutting out some of the most brilliant and interesting parts of the sky – such as the Southern Cross, the Centaur and so on, but short of taking the next aircraft to a more southerly latitude there is nothing to be done about it. The seasonal maps given here (Fig. 156) show only the principal stars. They cannot be very accurate, because trying to draw the sky on a flat piece of paper always presents problems, but they are much better than nothing, and they will at least indicate the directions in which to look.

Astronomers use the Latin names for the constellations: the Great Bear becomes Ursa Major, the Lion becomes Leo and so on. Also, each star is given a Greek letter followed by the genitive of the constellation name; thus Regulus is Alpha Leonis (Alpha of Leo), Polaris is Alpha Ursae Minoris (Alpha of the Little Bear), and Sirius is Alpha Canis Majoris (Alpha of the Great Dog). This was the scheme worked out in 1603 by a German astronomer

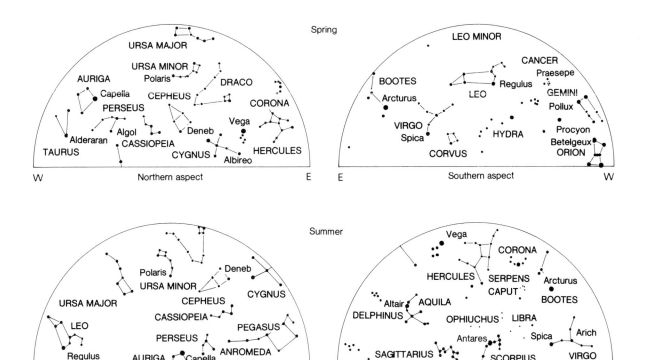

Spring

URSA MAJOR
URSA MINOR
Polaris
DRACO
AURIGA
Capella
CEPHEUS
CORONA
PERSEUS
Algol
Deneb
Vega
Alderaran
CASSIOPEIA
TAURUS
CYGNUS
HERCULES
Albireo

W   Northern aspect   E

LEO MINOR
CANCER
Praesepe
BOOTES
Regulus
GEMINI
Arcturus
LEO
Pollux
VIRGO
Procyon
Spica
HYDRA
Betelgeux
CORVUS
ORION

E   Southern aspect   W

Summer

Polaris
Deneb
URSA MINOR
CYGNUS
CEPHEUS
URSA MAJOR
CASSIOPEIA
LEO
PEGASUS
PERSEUS
Regulus
AURIGA   Capella   ANROMEDA

W   Northern aspect   E

Vega
CORONA
HERCULES
SERPENS
Arcturus
CAPUT
Altair   AQUILA
BOOTES
DELPHINUS
OPHIUCHUS   LIBRA
Antares   Spica   Arich
SAGITTARIUS   SCORPIUS   VIRGO

E   Southern aspect   W

156. Stars visible at the four seasons, during evenings.

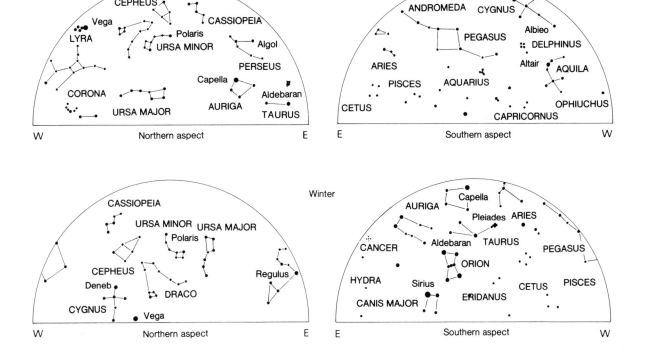

Autumn

CEPHEUS
Vega
CASSIOPEIA
LYRA
Polaris
Algol
URSA MINOR
PERSEUS
Capella
CORONA
Aldebaran
URSA MAJOR
AURIGA
TAURUS

W   Northern aspect   E

Deneb
ANDROMEDA   CYGNUS
Albieo
PEGASUS
DELPHINUS
ARIES
Altair   AQUILA
PISCES
AQUARIUS
CETUS
CAPRICORNUS
OPHIUCHUS

E   Southern aspect   W

Winter

CASSIOPEIA
URSA MINOR   URSA MAJOR
Polaris
CEPHEUS
Regulus
Deneb
DRACO
CYGNUS
Vega

W   Northern aspect   E

Capella
AURIGA
Pleiades   ARIES
CANCER
Aldebaran   TAURUS
PEGASUS
ORION
HYDRA   Sirius
CETUS   PISCES
CANIS MAJOR   ERIDANUS

E   Southern aspect   W

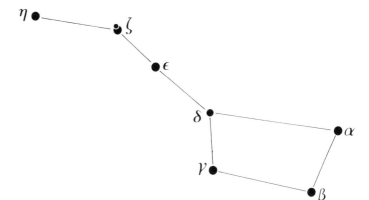

157. Ursa Major (the Great Bear or Plough).

named Bayer, and it has been followed ever since. If you want the Greek letters, they are given at the bottom of this page.* In theory, Alpha is the brightest star in a constellation, Beta the second brightest, and so on down to Omega. The order is by no means always followed, but it is some sort of a guide.

Individual names (most of them Arabic) are used for the brightest stars, but in general they are common only for the first-magnitude stars and a few special cases of stars which are fainter, such as the famous double star Mizar (Zeta Ursae Majoris) and the variable Algol (Beta Persei). In the maps there are also a few open and globular star clusters, some nebulae, and one galaxy, the Andromeda Spiral, which is the only external system clearly visible with the naked eye from Britain.

Chart 1. Circumpolar and Spring Constellations

Ursa Major (the Great Bear) is the most famous of all the northern constellations, and no part of it ever sets over the British Isles. Its seven chief stars make up the pattern known as the Plough or (in America) the Big Dipper. Even at its lowest it is still well above the horizon, and at times it may pass overhead. In Fig. 157 we have given a larger-scale map of it. None of its stars can claim to be brilliant, but all have proper names which are still often used: Alpha or Dubhe (magnitude 1.8), Beta or Merak (2.4), Gamma or Phad (also 2.4), Delta or Megrez (3.3), Epsilon or Alioth (1.8), Zeta or Mizar (2.1) and Eta or Alkaid (1.9). It is seen that Epsilon is the brightest of the Plough stars, though from its position in the Greek alphabet it should be only fifth. Any glance will show that Megrez is appreciably the faintest of the seven.

Ursa Major is useful as a direction-finder. Two of the Plough stars, Merak and Dubhe, show the way to Polaris or the Pole Star (2.0) in Ursa Minor (the Little Bear) which – as we have seen – lies within one degree of the celestial pole. Ursa Minor itself is not unlike a very faint and distorted version of Ursa Major, but it has only one other reasonably bright star, the orange Beta Ursae Minoris or

*Here is the Greek alphabet:

α	Alpha	ε	Epsilon	ι	Iota	ν	Nu	ρ	Rho	φ	Phi
β	Beta	ζ	Zeta	κ	Kappa	ξ	Xi	σ	Sigma	χ	Chi
γ	Gamma	η	Eta	λ	Lambda	ο	Omicron	τ	Tau	ψ	Psi
δ	Delta	θ	Theta	μ	Mu	π	Pi	υ	Upsilon	ω	Omega

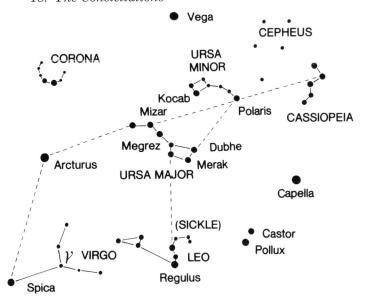

Chart 1. Circumpolar and Spring Constellations.

Kocab (2.1). When the sky is moonlit or misty, the other stars of the Little Bear will not be seen.

A line from Mizar through Polaris will lead on to another circumpolar constellation, Cassiopeia, whose five main stars make up a conspicuous W or M; the brightest of them are just below the second magnitude. Between Cassiopeia and Polaris lies the rather formless group of Cepheus, notable mainly because it contains a particularly important variable star, Delta Cephei, about which we will have more to say later.

Also circumpolar are two exceptionally brilliant stars, Vega in Lyra (0.0) and Capella in Auriga, the Charioteer (0.1). Though they never set, they skirt the horizon in the north when at their lowest. They lie on opposite sides of the Pole Star, and at roughly the same distance from it, so that when Vega is high up Capella is low down, and vice versa. It is fair to say that during summer evenings Vega is not far from the overhead point, while during winter evenings it is Capella which is not far from the zenith.

The other stars in Chart 1 are not circumpolar, but are well seen during spring evenings. Follow round the 'tail' of Ursa Major, and you will come to the brilliant orange star Arcturus, in Boötes (the Herdsman) (magnitude –0.04). The rest of Boötes is not spectacular, but not far away is the conspicuous semi-circle of stars marking Corona Borealis or the Northern Crown, with one second-magnitude star, Alpha Coronae or Alphekka. Further round, the curve from Ursa Major through Arcturus leads to Spica (1.0), leader of Virgo (the Virgin). Virgo is distinctive, and is shaped like a distorted Y; at the base of the Y is Gamma Virginis or Arich (2.7), a famous binary star. Adjoining Virgo is Leo (the Lion), with its leader Regulus (1.3); Leo can also be found by using the Pointers in the Plough 'the wrong way', i.e. away from Polaris instead of towards it. The curved line of stars which includes Regulus is known as the Sickle. Incidentally, remember that both Leo and Virgo are Zodiacal constellations, so that planets may sometimes be found in them.

Also on this chart you will see the Twins, Castor and Pollux, which link this chart with the Orion map (Chart 4).

149

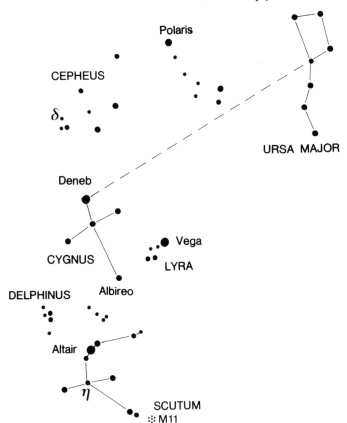

Chart 2. Summer Constellations.

Chart 2. Summer Constellations

During summer evenings, the scene is dominated by three brilliant stars, Vega in Lyra (the Lyre), Deneb in Cygnus (the Swan) and Altair in Aquila (the Eagle).*

Vega, as we have seen, is high up in summer after dark, and most people see it as blue in colour. Lyra itself contains no other bright star, though it has more than its fair share of interesting telescopic objects.

Cygnus (the Swan) is often known as the Northern Cross, but it is not symmetrical,

---

* Years ago, in a BBC *Sky at Night* programme, I nicknamed this pattern 'the Summer Triangle', and everyone now seems to use the name, though it is quite unofficial, and the three members of the Triangle are in different constellations! (P.M.)

since one member of the pattern – Beta Cygni or Albireo (3.1) – is fainter than the rest, and further away from the centre of the X, though to make up for this it proves to be a superb coloured double star when seen through a telescope. The leader of Cygnus is Deneb (1.2), a very luminous super-giant probably about 70,000 times as powerful as the Sun, and a very long way away from us.

The Milky Way flows through Cygnus, and the whole area is very rich. The Milky Way also passes through Aquila (the Eagle); the leader, Alpha Aquilae or Altair (0.8) has a fainter star to either side of it. Aquila has a distinctive shape; also in this region is the compact little constellation of Delphinus (the Dolphin), and lower down is Scutum (the Shield),

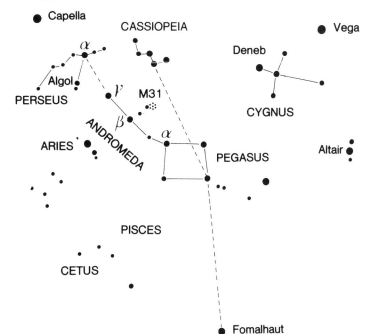

Chart 3. Autumn Constellations.

where the Milky Way is very striking, and where we find a lovely open cluster known popularly as the 'Wild Duck'. Its catalogue number is M11.

Deneb, like Vega, is just circumpolar from Britain, but Altair is not. Very low in the south you will also see Scorpius (the Scorpion); its leader, the red Antares (1.0) also has a fainter star to either side of it, but there is no danger of confusion with Altair, because Antares is so red and because it is much lower down. Scorpius itself is a magnificent constellation, but unfortunately it is never well seen from Britain, and part of it never rises at all. Adjoining it is Sagittarius (the Archer), with the star-clouds which indicate the direction of the centre of the Galaxy. Sagittarius has no first-magnitude star, and no obvious pattern, though many people nickname it 'the Teapot'.

Chart 3. Autumn Constellations

The key to the autumn chart is Cassiopeia, which, as we have noted, is circumpolar, and can be found by using Mizar and Polaris as pointers. Two of the stars in the W show the way to Pegasus, the Flying Horse of mythology, whose four chief stars make up a square; all are between magnitudes 2 and 3. Leading off from the Square comes the line of stars marking Andromeda.* Beyond Andromeda we come to Perseus, which has a shape which is easy to recognise but rather hard to describe. Its leader, Alpha Persei or Mirphak, is of magnitude 1.9, but the most celebrated star is Beta or Algol, the eclipsing binary which is usually just below the second magnitude.

* Alpheratz (2.1), in the Square of Pegasus, used to be known as Delta Pegasi, but has now been given a free transfer to Andromeda, and has become Alpha Andromedae. It is difficult to understand why, because it so clearly belongs to the Pegasus pattern.

Past Perseus, we come back once more to Capella.

Pegasus is high in the south during autumn evenings, but it is not so striking as might be thought from the map, and most people who do not know it expect it to be smaller and brighter than it really is. Well below it, past the chain of stars marking Pisces (the Fishes) is Fomalhaut in Piscis Australis (the Southern Fish), the southernmost of the first-magnitude stars ever visible from Britain. Even from South England it is always very low, and from North Scotland it barely rises at any time. Do not confuse it with the second-magnitude Beta Ceti or Diphda, in the constellation of the Whale, which is a magnitude fainter and is much higher up.

Chart 4. Winter Constellations

From winter through to spring Orion, the celestial Hunter, is dominant. It has two really brilliant stars, Beta or Rigel (0.1) and the orange-red Alpha or Betelgeux (variable, but usually around 0.5); Rigel is pure white, and is another 'searchlight', at least 60,000 times as luminous as the Sun. Betelgeux is much less luminous but

closer, so that the two stars appear similarly bright. The other stars in the main Orion pattern are of around the second magnitude. The proper names of the stars in Orion are often used. Mintaka, Alnilam and Alnitak make up the Belt; below the Belt is the misty Sword, containing the most famous of all gaseous nebulae, known as Messier 42. The celestial equator passes very close to Mintaka, the northernmost star of the Belt, so that Orion is visible at some time or other from every inhabited part of the world.

Orion's retinue is imposing. Upward, the Belt shows the way to the orange-red Aldebaran in Taurus (the Bull) and then to the lovely star-cluster of the Pleiades or Seven Sisters; downward, the Belt points to Sirius, which stands out at once. Sirius is pure white, but because of its brilliance and its low altitude it often seems to flash various colours of the rainbow. From Orion we can also find Procyon in Canis Minor, the Little Dog (magnitude 0.4) and the Twins in Gemini, Pollux (1.1) and Castor (1.6); the rest of Gemini is marked by lines of stars extending from Castor and Pollux in the

Chart 4. Winter Constellations.

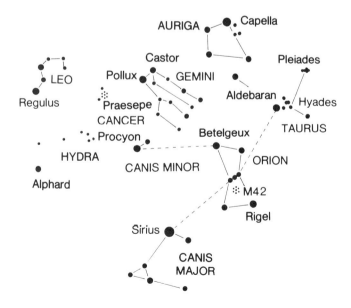

general direction of Betelgeux. Between the Twins and Leo lies the dim Zodiacal constellation of Cancer (the Crab), containing a famous open cluster, Praesepe or the Beehive; and the Twins point down to the reddish Alphard or Alpha Hydrae (2.0), the only bright star in Hydra, the Watersnake, which is actually the largest constellation in the whole of the sky.

We hope that these brief notes will enable you to make a start; the stars become so much more interesting when you know which is which.

Questions

1. Draw a map of Ursa Major, and show how it can be used to find Polaris, Cassiopeia and Arcturus.
2. Which are the two brightest stars in Orion, and why do they look so different?
3. You are observing the stars in Cancer, and find a bright object which is not on your map. What is it likely to be?
4. Draw a map of the Orion area, putting in Aldebaran, Castor, Pollux, Procyon and Sirius.
5. What is the main constellation high in the south during autumn evenings? How would you identify it?

Practical work

The best advice we can give here is to start by using the outline maps in this book, and then use more detailed maps to identify fainter and less obvious constellations. (A book list is given in the Bibliography.) Make sure that you can identify Orion, Ursa Major, Cassiopeia, Taurus, Gemini and Polaris, which are in the official GCSE list. You can also use a fixed camera to take star trails, and if you can mount your camera upon a driven telescope you can also take excellent pictures of entire constellations.

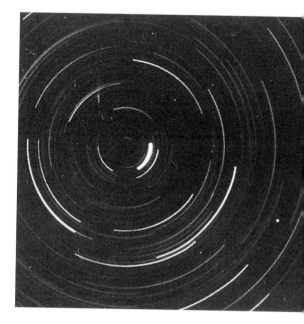

158. Star trails. Point your camera at the north celestial pole and give a time exposure; the result will be a series of star trails, as shown here. The short curved trail near the centre is that of Polaris itself – showing that Polaris is not exactly at the pole. This photograph was taken by Henry Brinton.

19
Introduction to the Stars

It is a long time now since astronomers realised that the stars are suns. Unfortunately, no telescope is capable of showing a star as anything but a point of light; if a star appears as a large, shimmering disk, you may be quite sure that your telescope is out of focus. The star should look like a sharp, well-defined point. And if the stars are as large as our Sun, they must clearly be a long way away.

There were other reasons, too, for supposing that all stars must be very remote; the lack of obvious change in the constellation patterns. Over a sufficiently long period the patterns will change, but the proper or individual movements of the stars are so slight that they have to be measured with very precise instruments. The holder of the 'speed record' is a faint red star, Barnard's Star, which has an annual proper motion of 10.3 seconds of arc per year against its background. This means that it would take around 190 years to cover a distance equal to the apparent diameter of the full moon.

William Herschel made a determined effort to measure star-distances by the method of *parallax*, but with no success, even though his observations led him to an important discovery – that of binary star systems, which we will discuss later. The first star-distance was measured in 1838 by the German astronomer F.W. Bessel. His method was quite straightforward in theory, though it was (and still is) difficult to apply in practice.

The principle of trigonometrical parallax is shown in Fig. 159, which is drawn out of scale for the sake of clarity. The Sun is represented by S, and E and F represent the Earth on opposite sides of its orbit. Since the Earth is 150,000,000 km from the Sun, the distance EF must be 300,000,000 km. X is a relatively nearby star, and is being seen against the background of more distant stars. When the Earth is at E, star X will be seen in position Y; six months later the Earth will have moved round to position F, and star X will be seen in the position Z. We can therefore measure the angle EXS (or FXS), which is the star's parallax. And since we know the distance ES (or FS), we can solve the whole triangle and work out the star's distance.

The main trouble is that the parallax shifts are so small – always less than one second of arc. Bessel selected the 5th-magnitude 61 Cygni, in the Swan, which is visible with the naked eye but has not been dignified with a proper name or even a Greek letter. Bessel chose it because it was a wide binary, and because it had an exceptionally large proper motion: 4.1 seconds of arc per year. He was able to measure the parallax, which amounted to about 0.3 seconds of arc, and hence to derive its distance.

The distance at which a star would subtend an annual parallax of 1 second of arc is 3.26 light-years, or rather over 31 million million km. This is known as the parsec (from parallax and second), and astronomers nowadays use it in preference to light-years. Bearing this in mind, we can do a simple mathematical calculation for 61 Cygni.

The modern value for the star's parallax is 0.292 seconds of arc. Divide 1 by

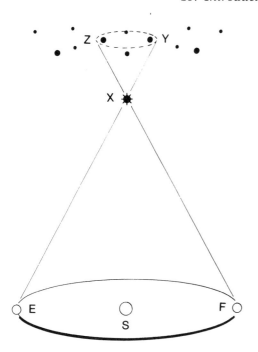

159. Trigonometrical parallax. Distance EF is known (it is the diameter of the Earth's orbit). From E, star X appears at Y; from F, X appears at Z. Therefore, the parallax angle EXS (or FXS) can be measured, and the distance found. The trouble is that except for relatively nearby stars, the parallax angles are so small that they are difficult to measure accurately.

0.292, and we have 3.424. This is the distance in parsecs. If you want to change it to light-years, multiply by 3.26. 3.424 x 3.26 = 11.16, which is the distance of 61 Cygni in light-years. Most of the stars are much further away than this, and beyond a distance of around 200 to 300 light-years the parallax shifts become too small to be measured with sufficient accuracy, so that we have to use less direct methods.

Shortly after Bessel announced his result, Thomas Henderson, a Scottish astronomer who had been working at the Cape of Good Hope, gave his measurement of the distance of Alpha Centauri, which is the third brightest star in the sky (it is too far south to be seen from Britain). Here the parallax is 0.76

seconds of arc. 1 divided by 0.76 is 1.31. This is the distance in parsecs. 1.31 x 3.26 = 4.3, the distance in light-years. Alpha Centauri is the closest of the bright stars; only its faint companion, Proxima, is nearer to us, by roughly one-tenth of a light-year.

Beyond the range of trigonometrical parallax, we have to work by finding out the star's real luminosity. Once this is known, together with the apparent brightness, we can work out the distance – after allowing for various complications such as the absorption of light by interstellar material.

We have already talked about *apparent magnitude*, which is the star's brightness as seen from Earth. It is a logarithmic scale. Two stars differing by 5 magnitudes differ in apparent brightness by a factor of 100. 1 magnitude is the fifth root of 100 ($5\sqrt{100}$), or about 2.512. Suppose, then, that we have two stars, one of magnitude 2 and the other of magnitude 12; what is their ratio of brightness? If a difference of 5 magnitudes indicates a ratio of 100, then with our two stars differing by 10 magnitudes the ratio must be 100 x 100, or 10,000. It is easy enough to see how the system works.

On the other hand, apparent magnitude is no clue to the star's real luminosity, because the distances are so different; a star may look bright either because it is close, or because it is genuinely very powerful, or a combination of both. For example, Sirius (apparent magnitude –1.5) is equal to 26 Suns put together, and appears much brighter than Rigel (+0.1) – but as we have seen, Rigel is some 60,000 times more luminous than the Sun. Sirius is a mere 8.6 light-years away, while Rigel lies at more like 900 light-years. Look at Sirius now, and you see it as it used to be 86 years ago, but Rigel is seen as it used to be at the time when William the Conqueror was drawing up the Domesday Book!

Absolute magnitude is defined as the apparent magnitude which a star would have if it could be seen from a distance of

10 parsecs (32.6 light-years). Sirius is closer than that; if taken out to 10 parsecs it would shine as a star of magnitude +1.4. On the other hand, from 10 parsecs Rigel would be superb. Its magnitude would be –7.1, and it would cast strong shadows.

There is a simple mathematical relationship linking apparent and absolute magnitude together with distance. If M = the star's absolute magnitude, m = the star's apparent magnitude, and d = the distance in parsecs, then: $M = 5 + m - 5 \log d$. Given two of these three values, you can work out the other. For example suppose a star 100 parsecs away has an apparent magnitude of 8; what is its absolute magnitude? Using the formula, we adapt it to the form $M = 5 + m - 5 \log d$. So $M = 5 + 8 - (5 \times 2) = 13 - 10 = +3$, the absolute magnitude of the star.*

---

* Many of you will not be familiar with the concept of logarithms (or 'logs') as they are no longer covered by GCSE. In truth, all that is necessary to use the formula is to press the appropriate calculator buttons, but for those who like a little more explanation a simple definition is given here. The logarithm of a number is simply the power to which 10 must be raised to give that number. 100 is 10^2 and so $\log(100) = 2$, 1000 is 10^3 and so $\log(1000) = 3$. (Strictly, this definition is only true for a certain kind of logarithm, those 'to the base 10' but that is all we need here.)

When stars are photographed, they leave their images on the sensitive plate or film; these images are of definite size, because of photographic effects. Generally speaking, the brighter the star, the larger its photographic image, and this gives a good way of obtaining its *photographic magnitude*. However, photographic magnitudes are not the same as visual magnitudes, because the original standard was worked out for a photographic system which was more sensitive to blue light than to red. Therefore, a blue star would give a bigger image than a red star which was equally bright to the eye. The difference between photographic magnitude and visual magnitude is the star's *colour index*; it is zero for white stars, negative for bluish stars, and positive for yellow or red stars. Thus a star of colour index –1 will be bluish, and a star with colour index +1.5 will be redder than a star whose colour index is, say, +1.1.

The colour of a star depends upon its surface temperature. And this brings us on to a discussion of the stars themselves – how they are born, how they evolve and how they die.

Questions

1. (a) Using a diagram, show what is meant by the trigonometrical parallax of a star.
   (b) What is the meaning of the term 'parsec'?
   (c) The star Procyon has an annual parallax of 0.287 seconds of arc. What is its distance in (i) parsecs, (ii) light-years?
2. (a) What is the brightness ratio between two stars A and B, whose apparent magnitudes are respectively 3.3 and 4.3?
   (b) A star of magnitude 2.1 lies at a distance of 10 parsecs. What is its absolute magnitude?
   (c) A star 100 parsecs away has an apparent magnitude of 7. What is its absolute magnitude?
3. (a) What is meant by the photographic magnitude of a star?
   (b) Define the term 'colour index'.
   (c) Two stars, A and B, have colour indices which are respectively +2.3 and –1.9. What does this tell you about the colours of those stars?

Practical work

1. Construct a model to show the principle of parallax. A good way to do this is to fix a ball on top of a pole, set it up in the garden, and then observe the ball's position against the background from opposite ends of a base-line.

2. Using the diagram of Ursa Major given on p. 148, make a model, showing the seven stars at their correct relative distances from a reference-point, and then align it with your eye so as to obtain the characteristic shape of the constellation. The values for the distances of the seven stars from the Sun (as given in the *Cambridge Catalogue 2000*) are as follows:

23 parsecs	Dubhe
19 parsecs	Merak
23 parsecs	Phad
20 parsecs	Megrez
19 parsecs	Alioth
18 parsecs	Mizar
33 parsecs	Alkaid

Then add Polaris (208 parsecs). Of all these stars, which do you think must be the most luminous?

20
From Young Stars to Black Holes

Because a star appears only as a point of light, telescopes can give us only a limited amount of information. They have to be used together with equipment based upon the principle of the spectroscope, which is all-important in the study of *astrophysics*, the physics of the stars. The problems are much more difficult than with the Sun, because there is never much light to spare – and even with modern electronic equipment, powerful telescopes are needed.

According to the modern system of classification, the stars are divided into various spectral types, each denoted by a letter of the alphabet. The sequence of letters is alphabetically chaotic – W, O, B, A, F, G, K, M, R, N and S – and many people know the famous mnemonic 'O Be A Fine Girl Kiss Me Right Now Sweetie',

though to make things even more difficult types R and N are now often combined into a single class, C. The main characteristics of these various types are as shown in the table.

Each type is subdivided into ten divisions; thus the Sun is of type G2, Rigel B8, Polaris F8, Aldebaran K5, and so on. The first two and the last three types are comparatively rare, and most of the stars have spectra of types B to M.

In the early part of our own century the work of two astronomers, E.J. Hertzsprung of Denmark and H.N. Russell of the United States, led to the drawing-up of diagrams now known as Hertzsprung-Russell or H-R diagrams, in which the stars are plotted according to their spectral types and their absolute magnitudes. (A 'colour-magnitude diagram',

Type	Surface temperature (°C)	Spectral characteristics	Colour
W	Up to 80,000	Many emission lines	White or bluish
O	40,000-35,000	Both bright and dark lines	White or bluish
B	30,000-12,000	Helium and hydrogen dominant	Bluish
A	11,000-7,500	Hydrogen dominant	White
F	7,500-6,000	Calcium lines prominent	Yellowish
G	6,000-4,200	Numerous metallic lines	Yellow
K	5,000-3,000	Strong metallic lines	Orange
M	3,400-3,000	Complicated; many lines due to molecules not simple atoms	Orange-red
R	2,600	Strong carbon lines	Red
N	2,500	Strong carbon lines	Red
S	2,600	Prominent bands of titanium oxide and zirconium oxide	Red

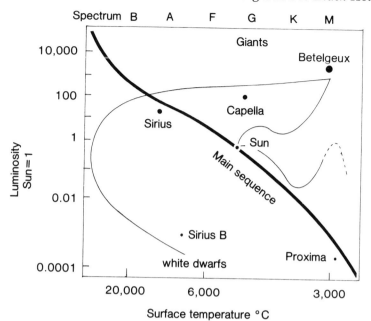

160. The H-R or Hertzsprung-Russell diagram.

obtained by plotting colour index against temperature, is of the same kind; in fact it comes to the same thing.) A typical H-R diagram is shown in Fig. 160, and one fact stands out at once: the distribution is not random. There is a well-marked belt running from the upper left down to the lower right; it is called the *Main Sequence*, in which we find the great majority of all stars. To the upper right we have the giant branch, while to the lower left we have the white dwarfs, which we know to be very old stars nearing the end of their careers.

Quite obviously, the yellow, orange and red stars are divided into two distinct classes, giants and dwarfs (we can ignore the white dwarfs for the moment). When the H-R diagrams were first published, it was tempting to regard them as an evolutionary sequence, with a star beginning as a large, cool red giant, shrinking to become a hot white star of type B or A, and then cooling down as it passed along the Main Sequence, becoming a dim red dwarf before fading away. Unfortunately things are not

nearly so straightforward as this, and the whole picture is wrong, even though we still often refer to 'early-type' stars (B and A) and 'late-type' stars (K and M).

It is true that a star begins its career by condensing out of the material in a nebula. As it shrinks, because of the force of gravity, it heats up. What happens next depends entirely upon its initial mass. If this mass is less than about eight times that of Jupiter, the most massive planet in the Solar System, it will never become a 'proper star', though its surface will become hot; objects of this type are known – rather misleadingly – as brown dwarfs, and during the past few years several have been identified with fair certainty, though we cannot yet pretend to know a great deal about them. And if the initial mass is less than about 0.1 that of the Sun, the core temperature will never rise high enough for nuclear reactions to be triggered off, and the star will shine as a dim red dwarf until it loses its energy. It may be said that a star of this type has failed its entrance examination!

If the mass is between 0.1 and 1.4

times that of the Sun, the whole story will be different. The 'protostar' will be cool and red (though this is not to say that it is a red giant, to the upper left of the H-R Diagram) and it will be varying irregularly; it has not settled down to a stable, steady existence. We can observe stars of this sort in nebulae such as the Sword of Orion. As they go on condensing, they heat up. When the core has passed the critical temperature of around 10,000,000°C, nuclear reactions start, and hydrogen is used as a 'fuel' in the way that I have already described when talking about the Sun. The star joins the Main Sequence, and remains there for a very long time. The Sun has been a Main Sequence star for around 5,000 million years, and it will be another 5,000 million years or so before its available hydrogen is used up and it has to change its structure drastically.

Helium, formed from the hydrogen, accumulates at the star's core. When there is no available hydrogen left, gravitation takes over again; there is further shrinking, which causes a new rise in temperature, and the helium starts to react, building up heavier elements such as carbon. From this point on, matters become very complicated, with new series of reactions leading to the building-up of heavier and heavier elements. The inside of the star shrinks, while the outer layers expand, so that the star leaves the Main Sequence and moves into the giant branch of the H-R Diagram. The core temperature rises to fantastic values, and the star throws off its outer layers altogether, turning into what is termed a *planetary nebula* – a bad term, because the object is not a true nebula, and has nothing whatsoever to do with a planet. But eventually, all nuclear reactions stop. The core, all that is left of the original star, is now inert. The star has become a stellar bankrupt – a white dwarf.

In a white dwarf, all the atoms are crushed and broken, and packed together

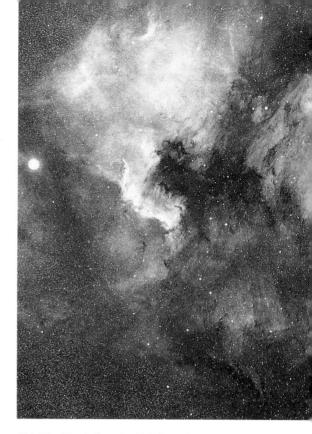

161. The North America Nebula in Cygnus, photographed in red light; photograph from Mount Wilson and Palomar Observatories.

so tightly that there is almost no waste space. This means that the density is very high, and may reach a million times that of water. A tablespoonful of white dwarf material might 'weigh' as much as a thousand tons. For example, Sirius has a white dwarf companion which is slightly more massive than the Sun, but is smaller than a planet such as Uranus or Neptune. Eventually, we may assume that a white dwarf loses the last of its energy, ending up as a cold, dead globe.

This must be the final fate of our Sun – but we will not be here to see. The Earth cannot survive the period when the Sun is a red giant, at least a hundred times as luminous as it is today. It is a depressing thought, but at least we are in no danger at the moment!

If the original star is more than 1.4 times as massive as the Sun, everything will happen at a quicker pace. The star will condense out of a nebula, as before, and join the Main Sequence, but the greater mass means that when the hydrogen 'fuel' is exhausted still heavier elements are produced, and the temperature rises to something like 3,000,000,000°C. By now the core is composed mainly of iron, which will not react. Suddenly the nuclear processes stop; there is an 'implosion', followed by a rebound and an explosion, and the star blows up in a cataclysmic outburst which we call a supernova.* Much of the material is hurled away into space, and the end product is a cloud of expanding gas in the midst of which is a very small, super-dense object made up of neutrons. The atomic protons (positively charged) and electrons (negatively charged) are forced together to produce neutrons, which have no electrical charge at all.

A neutron star is only a few kilometres across, but its density may be of the order of a thousand million million tons per cubic centimetre. The star has an immensely powerful magnetic field, and is spinning round rapidly – often many times per second. It is sending out radio pulses in two opposite directions, and every time a pulse sweeps across the Earth, rather in the manner of a rotating lighthouse beam, we receive a radio signal, which is why neutron stars are often called *pulsars*. When the first pulsar was discovered, in 1967, the signals were so rhythmical and so peculiar that for a few days it was even thought that they might be artificial. The most famous pulsar is inside the patch of gas we know as the Crab Nebula, in Taurus; it is known to be the remnant of a supernova which was seen in the year 1054, and for a while became bright enough to be visible with the naked eye in broad daylight.

As a pulsar spins, it loses energy, and slows down – very gradually, but by a

measurable amount. Eventually it, too, will become cold and dead.

If the initial mass of a star is still greater – at least 8 times that of the Sun – an even stranger fate will overtake it. When the final collapse starts, it is so sudden and so violent that nothing can stop it. There cannot even be a supernova outburst; the star simply goes on collapsing and collapsing, and becoming denser and denser. As it does so, the escape velocity rises. There must come a time when the escape velocity has reached 300,000 km per second. This is the velocity of light, so that light can no longer escape from the star – and if light cannot do so, then certainly nothing else can, because light is the fastest thing in the universe. The old star has surrounded itself with a 'forbidden area' from which nothing can break free. It has become a black hole.

We cannot see a black hole, because it emits no radiation, and all we can do is to track it down by its effects upon objects which we can see. One good candidate is the system known as Cygnus X-1, about 6500 light-years from us. Here we have a blue supergiant star associated with an

162. The Orion Nebula; Hubble Space Telescope. The picture is composed from 45 separate fields.

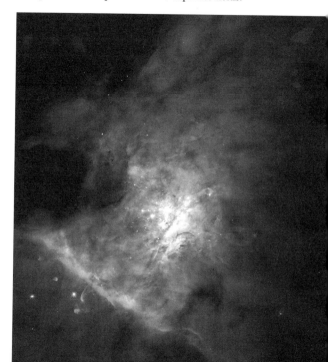

* This is a so-called Type II supernova. There is another kind of supernova, as we shall see.

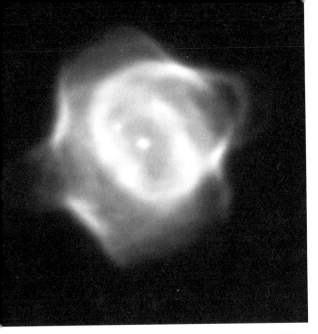

163. The Stingray planetary nebula (Henize 1357); Hubble Space Telescope.

invisible companion which seems to be about 14 times as massive as the Sun, as

we can tell from its gravitational effects on the supergiant. Material from the supergiant is being pulled into the black hole, and before being sucked in it is so strongly heated that it gives off X-rays which we can detect (which accounts for the name Cygnus X-1).

Black holes are the most bizarre objects we can imagine – indeed, it is very difficult to imagine what they must be like. This is no place to go into detail, and there is a great deal that we do not know, but the evidence seems to indicate that black holes really do exist.

At least we have one consolation. The Sun is a normal Main Sequence star, and it has a long period of steady existence before anything much happens to it. It is not massive enough to become a supernova or a black hole, and there is no reason to think that it will destroy our world for at least another five thousand million years in the future.

Questions

1. Construct a Hertzsprung-Russell diagram, putting in the stars given in the table.

Star	Spectral type	Absolute magnitude
The Sun	G2	+4.8
Sirius	A1	+1.4
Rigel	B8	–7.1
Betelgeux	M2	–5.6
Companion of Sirius	A	+11.4
Barnard's Star	M5	+13.2
Procyon	F5	+2.7
Spica	B1	–3.5
Fomalhaut	A3	+2.0
Polaris	F8	–4.6
Aldebaran	K5	–0.3
Vega	A0	+0.5
Antares	M1	–4.7
Canopus	F0	–8.5
Wolf 359	M8	+16.7
Altair	A7	+2.2
Capella	G8	+0.3

2. What is a pulsar? Why is its density so high?

3. What would you take to be the colours of the following stars: (a) Regulus (spectral type B7), (b) Pollux (K0), (c) Arcturus (K2), (d) Alkaid (B3), (e) Chi Cygni (S)?

4. In what way can we hope to detect a black hole?

5. Why can we be confident that our Sun will not suffer a supernova outburst?

21
Double Stars, Variable Stars and Novae

The Sun is a single star, but in the sky we find many examples of stars which are double or even multiple. Not all these are true pairs. Consider stars A and B in Fig. 164; they are very distant from each other, but from the Earth they lie in almost the same direction, so that in the sky they look side by side. This makes up what is called an *optical* double.

Now look at Mizar, the second star in the tail of the Great Bear (Fig. 165). Close beside it you will see a much fainter star, Alcor, which is of the fourth magnitude. Use a telescope, and you will see that Mizar itself is double, with one component considerably brighter than the other; but the separation is less than 15 seconds of arc, so that with the naked eye they appear as one star. They are genuinely associated, and make up what is termed a *binary* system.

The stars in a binary system move together round their common centre of gravity. We have come across this sort of orbital motion before – remember how the

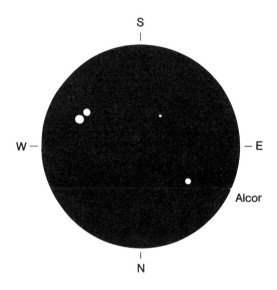

165. Mizar; it is a binary, and in the same field with Alcor. The star between the Mizar pair and Alcor is not connected with the system.

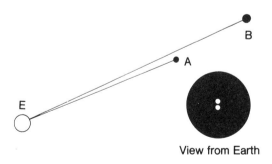

View from Earth

164. Optical double. From the Earth (E) the stars A and B will appear very close together, though in fact they are far apart.

Earth and Moon revolve round the barycentre – but with binaries, the whole scale is different. Though the stars are so unequal in size and luminosity, they are not nearly so unequal in mass, because the smaller stars are always the denser; it is rather like balancing a meringue against a lead pellet. If the components are of equal mass, the barycentre will be midway between them, while if the components are unequal the barycentre will be displaced toward the more massive component (Fig. 166). Look, for example, at Sirius, which is 26 times as luminous as the Sun and has approximately 2½ times the solar mass. The tiny white dwarf companion has a diameter

164

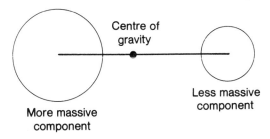

166. Centre of gravity of a binary system with components of unequal mass.

only three times greater than that of the Earth, but a mass a little more than that of the Sun, so that their common centre of gravity is not so displaced as might be thought. The revolution period is 50 years (Fig. 167). The companion is of magnitude 8.5, and if it could be seen on its own it would be visible with binoculars, but it is so drowned by the brilliance of Sirius itself that it is decidedly elusive, particularly as the apparent separation between the two is less than five seconds of arc.

Binary stars are surprisingly frequent – more so, in fact, than optical pairs. In such a system the components must have a common origin, but they can be very wide apart, as in the case of Mizar, where the orbital motion is so slow that all we can really say is that the two stars share a common motion through space.

Many binary stars are within the range of a small telescope, and we also have triple and multiple pairs. A particularly good example is Epsilon Lyrae, not far from the brilliant Vega (Fig. 168). Keen-sighted people can see that it is double; both components are of about the fifth magnitude. A 3-inch (7.6-cm) telescope is enough to show that each component is again double, so that we have a quadruple system. Between the two pairs is another star, which lies in the background and is not associated with Epsilon Lyrae in any way.

Some pairs have contrasting colours. Pride of place must go to Albireo, in Cygnus, where the third-magnitude primary is yellow and the fifth-magnitude companion blue.

If the components of a binary are very close together, no telescope will show them separately, but – as so often – we can make use of the spectroscope. Fig. 169 shows how it is possible to observe a *spectroscopic binary*. We have already referred to the Doppler effect, which means that an approaching star will seem a little too blue, while a receding star will seem a little too red. The effect shows up in the spectrum. With a star which is coming toward us, all the dark lines will be blue-shifted, while with a star which is moving away from us the lines will be

167. Orbit of the companion of Sirius. Its revolution period is 50 years. Sirius A is 10,000 times as luminous as its companion, but only 2½ times more massive.

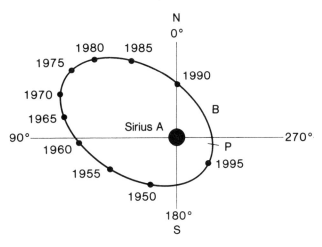

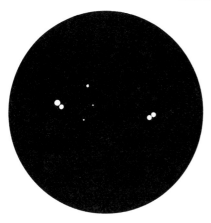

168. Telescopic view of the double-double star Epsilon Lyrae. The other three stars are not associated with it.

shifted to the long-wave or red end of the spectrum. The greater the velocity, the greater the shift.

In Fig. 169 we have two stars, A and B, moving round their common centre of gravity. In position 1, A is approaching us, and the lines in its spectrum will be blue-shifted; B is receding, and the lines will be shifted to the red. In position 2, the reverse applies, with A receding and B approaching. By watching the behaviour of the lines, we can work out the nature of the system. Even if one component is so faint that its spectrum cannot be seen, the oscillations in the lines of the visible component will betray the fact that we are dealing with a binary pair. (Of course, these effects are superimposed on the overall *radial velocity* of the whole system towards or away from us, but the principle is clear enough.)

Some stars are remarkably complex. Consider Castor, the senior though fainter of the two Twins. Telescopically it is a wide, easy double; each component is a spectroscopic binary, and there is a third, much fainter component which is also a spectroscopic binary, so that altogether Castor is made up of six suns, four brilliant and two dim.

Most stars shine steadily over very long

periods, but there are some which do not; instead, they brighten and fade over short periods of time – sometimes regularly, sometimes irregularly. These are the *variable stars*, to which we must now turn.

The way to estimate the magnitude of a variable star is to compare it with two neighbouring stars which do not change, and whose magnitude you know. For example, suppose we have two stars, A and B, whose magnitudes are respectively 4.2 and 4.6, and estimate that the variable is exactly half-way between them; what is its magnitude? Clearly, the answer must be 4.4. Things are not usually as easy as this, but we have given more details in the Practical Work Section at the end of this chapter. Meantime, what sorts of variable stars can we find?

Oddly enough, our first examples are not truly variable at all, even though they show changes in light. They are known as *eclipsing variables* or *eclipsing binaries*, of which the most celebrated example is Algol, in Perseus. There are two components, one considerably brighter than the other, completing one revolution

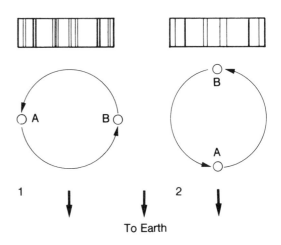

169. A spectroscopic binary. In position 1, lines from A are blue-shifted, lines from B red-shifted. The reverse applies in position 2.

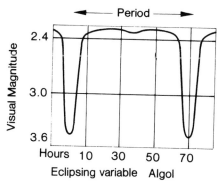

170. Light-curve of the eclipsing binary Algol.

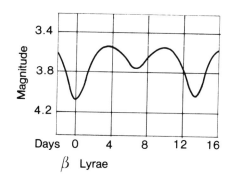

171. Light-curve of Beta Lyrae.

round their common centre of gravity in 2½ days, and therefore much too close together to be seen separately. The angle from which we view the system is such that the fainter star regularly passes in front of the brighter, cutting out part of its light, so that every 2½ days Algol seems to 'wink'; it takes four hours to fade from magnitude 2.2 to 3.4, remaining at minimum for only twenty minutes before slowly brightening again. (With Algol, the eclipse is not total, but the principle is clear enough.) This is shown in Fig. 170, together with a *light-curve* in which magnitude is plotted against time. There is a small secondary minimum when the fainter star is eclipsed by the brighter, but in Algol's case the drop in magnitude is too slight to be noticed with the naked eye; you have to use a measuring device known as a photometer, which need not concern us at the moment.

A few eclipsing binaries are bright enough to be followed with the naked eye. One of these is Beta Lyrae, near Vega, where the components are less unequal, so that there are alternate deep and shallow minima (Fig. 171) – and with Epsilon Aurigae, one of a triangle of stars close to Capella, eclipses happen only once every 27 years.

With genuine variables, the light-changes are due to real alterations in the star's output. Among the most important are the *Cepheids*, so named because the

best-known member of the class is Delta Cephei, in the far north of the sky (Fig. 172). The range is from magnitude 3.5 to 4.4, so that the star is always easily visible with the naked eye. Its period, or interval between one maximum and the next, is 5.3 days, and this is absolutely regular, so that we always know what magnitude the star will be at any particular moment. The light-curve is shown in Fig. 173. The curve is repeated time and time again, with clockwork precision.

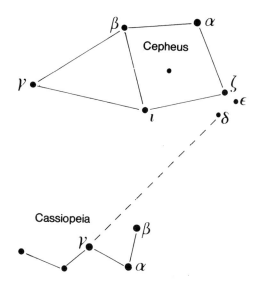

172. Delta Cephei.

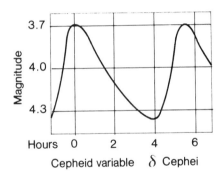

173. Light-curve of Delta Cephei.

Cepheids are giant stars, and are common in the Galaxy. What makes them so important is that their periods are linked with their real luminosities; that is to say, a Cepheid with a period of 5.3 days will have the same luminosity, in terms of the Sun, as any other Cepheid with a 5.3-day period. The longer the period, the greater the luminosity: thus Eta Aquilae in the Eagle, a Cepheid with a period of 7.2 days, is more powerful than Delta Cephei itself. Once we know the real luminosity, as well as the apparent magnitude, we can work out the distance.

As an example, let us go back to Delta Cephei and Eta Aquilae. The apparent magnitudes of the two are exactly the same, but since Eta Aquilae is the more luminous it must also be the more remote. Cepheids act as our 'standard candles', and have been of immense value, because they are seen in other galaxies as well as our own. They are old stars, usually of spectral types F and G, and they are pulsating, so that they swell and shrink regularly. Associated with them are pulsating variables of much shorter period, known as RR Lyrae stars, all of which seem to be about 90 times as luminous as the Sun, and which can therefore be used in the same way.

Long-period variables, now generally called Mira stars after the brightest member of the type (Mira or Omicron Ceti, in the Whale) behave differently.

They are red giants of types M or later; their magnitude ranges are much greater than for the Cepheids, and their periods are longer, ranging from a few weeks to well over a year. Mira itself (Fig. 174) has a period of 332 days. At its brightest it may reach the second magnitude, and occasionally outshines the Pole Star, but at minimum it drops to magnitude 10, so that a telescope is needed to show it. Both the period and the range fluctuate to some extent, and there is no direct link between period and luminosity, as with the Cepheids (Fig. 173).

Mira stars are very common, but not many become bright enough to be seen with the naked eye even at maximum. However, Chi Cygni, in the Swan, may reach magnitude 3.3. At minimum it becomes very faint – below magnitude 14 – and the period is 407 days.

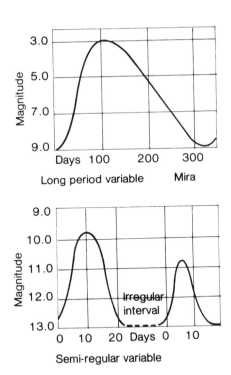

174. Light-curves of Mira and a semi-regular variable. (With semi-regular variables, the periods are very rough.)

Semi-regular variables also pulsate, but they have much smaller ranges than the Mira stars, and both the periods and the amplitudes are themselves variable. Betelgeux in Orion comes into this category. At times it may rival Rigel, while at other times it is little brighter than Aldebaran. There is a very rough period of about five years, but the changes are always slow.

Of the *irregular variables*, perhaps the most interesting are those of the R Coronae class. R Coronae itself, in the 'bowl' of the Crown, is usually just below the sixth magnitude, so that it is on the fringe of naked-eye visibility and is easy with binoculars, but at unpredictable intervals it drops to minimum, falling to perhaps the 15th magnitude and taking some time to recover. Apparently it contains more than the usual amount of carbon, and when this carbon accumulates in its atmosphere the star literally hides itself behind a veil of soot.

The most spectacular of all variable stars are the novae. 'Nova' is Latin for 'new', but a nova is not a new star at all; what happens is that a formerly faint star suddenly flares up, remaining bright for a few days, weeks or months before fading away to its former magnitude. In 1918 a nova in Aquila reached magnitude –1.1, so that it outshone every star in the sky apart from Sirius, but it has now become a very dim telescopic object. The last really bright nova was seen in Cygnus, in 1975. It reached magnitude 1.8, but within ten days it had become invisible with the naked eye.

A nova is a binary system, with a low-density Main Sequence star together with a white dwarf companion. The white dwarf pulls material away from the primary, and a ring of material is built up around the white dwarf; when enough of it has accumulated, there is a nuclear outburst, and gas is ejected at high velocity. At the end of the outburst, the system returns to its old state. A few stars, such as T Coronae in the Crown, have been known to show more than one outburst (with T Coronae, in 1866 and again in 1946); these are called *recurrent novae*.

Do not confuse a nova with a supernova, which is an outburst on a much larger scale. As we have seen, a Type II supernova marks the death of a very massive star. A Type I supernova is different. Here we have a binary system, but when the outburst happens the white dwarf companion is completely destroyed, producing a blaze of radiation which may be equal to a thousand million Suns put together. During the past thousand years only four supernovae have been definitely seen in our Galaxy – those of 1006, 1054, 1572 and 1604 – though in 1987 a Type II supernova appeared in a neighbouring galaxy, the Large Magellanic Cloud, at a distance of 170,000 light-years. The 1054 supernova, described by Chinese astronomers, has left the remnant which we know as the Crab Nebula. Like many other supernova remnants, it is a powerful source of radio radiation.

Variable star observation has become one of the most important branches of amateur astronomy, if only because there are so many variables that professional workers cannot begin to cope with them all. If you want to carry out some really useful research as well as taking your GCSE, variable stars will present you with plenty of opportunities.

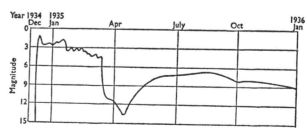

175. Light curve of the nova DQ Herculis, 1934 – a slow nova, which showed a temporary revival before its final fading.

Questions

1. (a) What is the difference between an optical double star and a binary?
 (b) Sirius is only about 2½ times as massive as its companion, but is 10,000 times as brilliant. Explain this apparent discrepancy.
 (c) What are spectroscopic binaries, and how are they identified?
2. (a) Using a diagram, explain why Algol changes in magnitude even though it is not intrinsically variable.
 (b) A and B are stars with magnitudes of 2.6 and 3.8 respectively. A variable star is seen to be midway in brightness between these two. What is the magnitude of the variable?
 (c) What colour are most of the Mira variables?
3. (a) Explain why Cepheid variables are so important to astronomers.
 (b) Eta Aquilae and Zeta Geminorum are both Cepheid variables. Eta Aquilae has a period of 7.2 days, while the period of Zeta Geminorum is 10.1 days. Their apparent magnitudes are almost the same. Which star is the more remote, and why? (You may ignore factors such as the absorption of light in space.)
 (c) Why does R Coronae show sudden, unpredictable drops in brightness?
4. (a) Explain the cause of a nova outburst.
 (b) What is the difference between a nova and a supernova?
 (c) Why is amateur observation of variable stars of real scientific importance?

Practical work

1. *Double stars*. A recommended project is in observing double stars, noting their colours, magnitudes and separations. Almost all well-known double stars are binary systems. For most of these you will need a telescope, but some are within range of binoculars or even the naked eye, and I have given a few typical examples:

(a) Mizar (Zeta Ursae Majoris). This has already been described. The magnitudes are 2.3 and 4.0; separation 14.4 seconds of arc, so that a small telescope is adequate. Both components are white. With a low-power eyepiece, Alcor is also in the field of view. The star between Alcor and the Mizar pair is unconnected with the system.

(b) Epsilon Lyrae, near Vega, also described in the text. Both components can be separated with a 3-inch (7.6-cm) refractor, though not easily, as in each case the separation is less than 3 seconds of arc.

(c) Beta Cygni (Albireo), in the Swan. Magnitudes 3.4 and 5.1; separation 34.4 seconds of arc, so that this is a very easy pair. The primary is a yellow star, the companion blue.

(d) Gamma Virginis or Arich. The components are equal at magnitude 3.5. The present separation is 3 seconds of arc, so that this is an easy pair, but Arich is a binary with a period of 171 years, and the separation is becoming less – not because the components are really getting closer together, but because we are seeing them from a narrower angle. By the year 2016 the star will appear single except in large telescopes.

(e) Polaris. This is a more difficult double. The Pole Star itself is of magnitude 2.0, the companion 9.0; the separation is 18.4 seconds of arc, but the faintness of the companion makes it hard to see with any telescope of much less than 4 inches aperture.

2. *Variable stars.* As I have already said, this has become a very important branch of amateur astronomy. Eye estimates, using suitable comparison stars, can be accurate to a tenth of a magnitude. (If you look at Orion's Belt, you will see that Epsilon or Alnilam, the middle star, is just brighter than Zeta or Alnitak, the lower star, but the difference is less than 0.1 magnitude – Alnilam is +1.70, Alnitak +1.77. If you want to be more accurate, you will have to use a photometer, but this is not for the beginner.

Choose at least two comparison stars – three if possible. Sometimes you may find that things do not quite 'fit', and a compromise is needed. For example, suppose there are three comparison stars: A (4.0), B (5.0) and C (4.6). You make the variable 0.4 fainter than A, 0.5 brighter than B and 0.2 brighter than C. From A and C, the magnitude of the variable works out at 4.4; from B, 4.5. Re-check, and if there is still a discrepancy take the better value, which would here be 4.4. Also, many variables (particularly Mira stars) are red, and it is not too easy to compare a red star with a white one.

Train yourself to identify differences of one-tenth of a magnitude, and work accordingly. There are other methods, but the simple procedure will do for the moment (we always use it in preference to the other, more complicated methods, but everyone will have his own ideas).

It may help to give a selection of variables for observation, but of course there are many others – you have a wide choice.

(a) *Algol* (Beta Persei) (Fig. 176). The eclipsing binary. The times of minima can be looked up, and the variations are quite noticeable over short periods, so that a light-curve can be drawn up. The available comparison stars are Gamma Andromedae (2.1), Delta Cassiopeiae (2.7), Zeta Persei (2.9) and Epsilon Cassiopeiae (3.4), all of which are shown on the chart. Avoid Rho Persei, which is a semi-regular variable with a range of from 3.4 to 4 and a very rough period of a few weeks.

(b) *Betelgeux* (Alpha Orionis). We include this only because it is so bright. Compare it with Aldebaran (0.8), Procyon (0.4) and occasionally Rigel (0.1), but there are difficulties because the lower down a star is, the more its light will be dimmed ('extinction'). Thus if Betelgeux and Aldebaran look equal, but Betelgeux is lower down over the horizon, then it will actually be the brighter of the two.

(c) *Mira* (Fig. 177). The range is from around 3 to 10, with a period of 332 days. Mira is visible with the naked eye for only a few weeks in each year, but its maxima can be predicted to within a week or two. If you see it, you can use Alpha Ceti (2.5), Delta Ceti (4.1), Zeta Ceti (3.7) and Alpha Piscium (3.8) as comparisons.

(d) *Scheat* (Beta Pegasi) (Fig. 178). This is one of the stars of the Square of Pegasus. The range is from 2.4 to 2.9, with a rough period of around 38 days, so that it belongs to the semi-regular class. Use Alpha Pegasi or Markab (2.5) and Gamma Pegasi or Algenib (2.8), but beware of extinction when using Algenib. Scheat is obviously orange (spectrum M2).

(e) *Eta Aquilae* (Fig. 179). A typical Cepheid; 3.5 to 4.4, 7.2 days. I have selected it in preference to Delta Cephei itself because it is easier to find; it lies between Theta Aquilae (3.3) and Delta Aquilae (3.4). Also use Iota Aquilae (4.4).

All these variables (a to e) can be followed with the naked eye.

(f) *Rho Cassiopeiae*, near the W of Cassiopeia (Fig. 176). Use binoculars. We have included it because nobody is sure what kind of variable it is, and therefore it is worth watching. Usually it is around magnitude 5, but occasionally drops to below 6. The best comparison stars are Sigma (4.9) and Tau (5.1). Rho Cassiopeiae is very remote and luminous; it is an F-type supergiant.

(g) *R Lyrae*, near Vega (Fig. 180). 4.0 to 5.0; semi-regular, period about 46 days. Use binoculars with a wide field, as the best comparison stars, Theta and Eta Lyrae, are rather

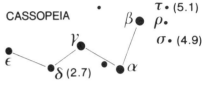

CASSOPEIA

176. Algol, Rho Persei and Rho Cassiopeiae.

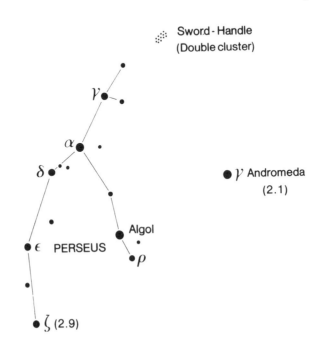

Sword - Handle
(Double cluster)

γ Andromeda
(2.1)

Algol

PERSEUS

177. Mira.

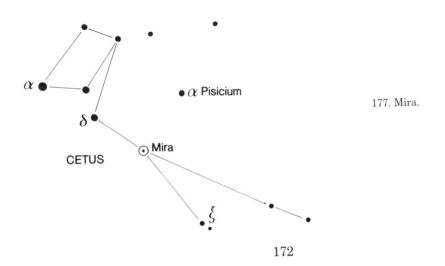

α Pisicium

CETUS

Mira

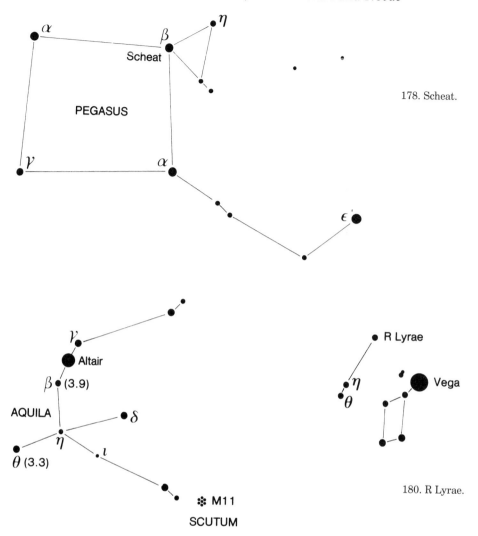

178. Scheat.

179. Eta Aquilae.

180. R Lyrae.

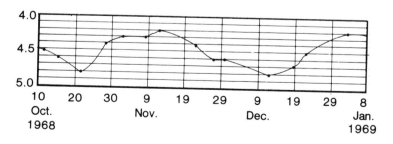

181. Light-curve of R Lyrae, from P.M.'s binocular estimates.

a long way from it. Both these are listed as 4.4, but I make Theta about a tenth of a magnitude the brighter of the two. A light-curve of R Lyrae over one cycle is given here (Fig. 181). It was drawn from estimates, and at least it shows how the star behaves.

(h) *R Coronae Borealis* (Fig. 182). In the 'bowl' of the Crown. Binoculars generally show it as of around magnitude 6; compare with the other binoculars star in the Crown, which is 6.6. As we have seen, R Coronae sometimes falls to a very faint minimum. If you look for it and cannot find it, you may be sure that this is what is happening.

(i) *R Cygni* (Fig. 183). We have included this 'for serious observers only'. It has a period of about 426 days, but as with all Mira stars this period is not quite constant. It is never visible with the naked eye, but binoculars will show it when it is at maximum. It lies very near the fourth-magnitude Theta Cygni, and is therefore easy to find. Locate Theta Cygni, and turn your telescope toward it. You will see the star field given in Fig. 184 (opposite). This field has been drawn to cover 1 square degree, which means using a fairly low magnification to see the whole field at once. The star lettered 2 is on one side of Theta; R is on the other.

If R is near minimum, you will not see it unless you have a fairly large telescope. (With a 12½-inch reflector it can be followed through its entire cycle, but not easily.) If you want to study it, start looking for it when it is near maximum. Having located it, you will be able to follow it until it has faded below the limiting magnitude of your telescope.

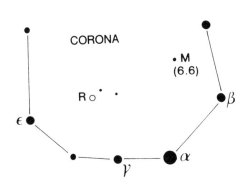

182. R Coronae Borealis.

183. Finding R Cygni. The first step is to locate
Theta Cygni, magnitude 4.6. R Cygni is in the same
telescopic field.

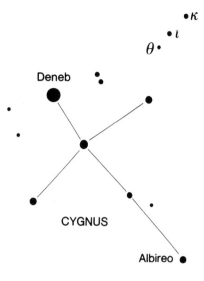

CYGNUS

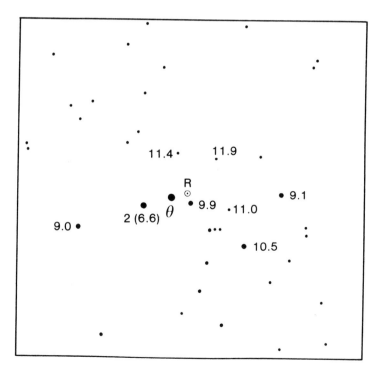

184. Field of R Cygni. This
covers 1 square degree. The star
marked 2 (magnitude 6.6) is to
one side of Theta: R is to the
other. The other most-used
companion stars are shown with
their magnitudes. Thus the star
close to R, rather below it on the
chart, is of magnitude 9.9.

22
The Milky Way

As well as its hundred thousand million stars, the Galaxy contains many other objects, in particular the star-clusters and nebulae. Over a hundred of these were listed by the French astronomer Charles Messier in 1781 – not because he was interested in them, but because he kept on confusing them with new comets. We still use the Messier or M numbers, though astronomers prefer what are known as the NGC numbers – those of the 'New General Catalogue', now over a hundred years old. Many now use the numbers in the Caldwell catalogue, which lists 109 light clusters and nebulae not listed by Messier.

Star clusters are exactly what their name suggests: groups of stars much more closely packed than in average regions, though they are still too widely spread to be in danger of collision. The best-known *open cluster* is the Pleiades (M.45), otherwise known as the Seven Sisters, in Taurus (the Bull). The brightest member

is Alcyone, of the third magnitude. Normal-sighted people can see at least seven stars with the naked eye, making up a compact little swarm that cannot be overlooked; some people can see at least a dozen, and the record is said to be nineteen. Binoculars will show many more, and altogether the cluster contains several hundred stars. The leading stars of the Pleiades are hot and white; the cluster is young, and photographs show a good deal of nebulosity, so that star formation is still going on. There is every reason to believe that the cluster stars had a common origin (Fig. 187).

Among other open clusters are the Hyades, round Aldebaran, and Praesepe in Cancer, between the Twins and Leo. The Hyades are very scattered, so that binoculars actually show them better than a telescope. They are overpowered by the bright orange light of Aldebaran, but in fact Aldebaran is not a genuine

185. Composite picture of the Milky Way.

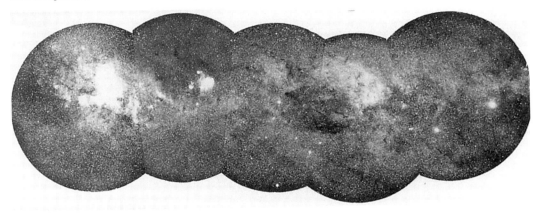

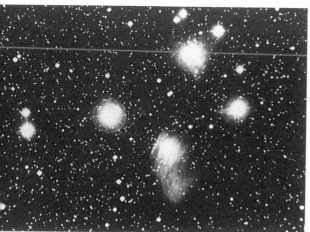

186. The Pleiades star-cluster (M.45), often called the Seven Sisters. In this picture the associated nebulosity can be seen.

member of the cluster; it merely happens to lie about midway between the Hyades and ourselves. M.44, Praesepe or the 'Beehive', is an easy naked-eye object, though moonlight will drown it. Look also for the twin clusters in the 'sword' of Perseus, which for some reason are not in Messier's list. They are magnificent objects when seen through a low-power telescope.

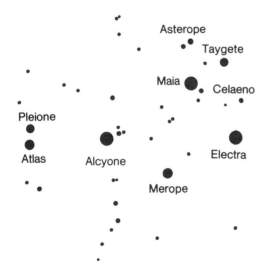

187. The Pleiades.

Plenty of open clusters are within binocular range – some rich, some sparse. They are recognisable at a glance, but this is not so with the *moving clusters*, which are made up of stars which share a common motion through space even though they may look far apart. Five of the seven stars in the Plough make up a moving cluster (the non-members being Alkaid and Dubhe), and many others are known.

Quite different are the *globular clusters* – huge, regular systems of stars. A typical globular may contain up to a million stars, arranged in a spherical form; near the centre of the clusters the stars seem so closely packed that they merge into a glowing mass as seen through a telescope. The distances between the stars in these regions are only light-days or light-weeks. If our Sun were a member of a globular cluster, the night-sky would be superb; there would be many stars bright enough to cast shadows, and there would be no true darkness at all. Moreover, many of the stars would be red, because globular clusters are very old, and their leading members have evolved off the Main Sequence and have become giants.

All the globular clusters are a long way away. Few are within 20,000 light-years of us, and only just over one hundred have been found in the entire Galaxy. From Britain, the only one visible with the naked eye is M.13, in Hercules. It is far from bright, and difficult to find without optical aid unless its position is known, but binoculars show it clearly, and telescopically it is a splendid sight. In the far south of the sky there are two brighter globulars (Omega Centauri and 47 Tucanae), but unfortunately both remain permanently below the British horizon.

The globular clusters lie round the edges of the main Galaxy. Their distances were first measured when they were found to contain short-period variable stars – not true Cepheids, but RR Lyrae stars, which have periods of only a day or so and all of which are about 90 times as luminous as the Sun. Once the periods and magnitudes of the RR Lyrae variables had been worked out, their distances could be found, and this enabled the American astronomer Harlow Shapley to find the distances of the globulars themselves.

This may be the moment to mention what are called 'stellar populations'. In regions of Population I, the brightest stars are hot, white and relatively young; in Population II regions, the brightest stars are old red giants and supergiants. Globular clusters, then, are of Population II.

When Shapley carried out his research, during and just after the First World War, he found that the globular clusters are not spread evenly all over the sky. They are concentrated in the southern hemisphere, particularly in the area of Sagittarius, the Archer. Shapley realised that this is because we are having a lop-sided view of the Galaxy; the Sun lies

188. The globular cluster M.13 in Hercules, the brightest such object in the northern hemisphere of the sky. It is 22,500 light-years away, and is just visible with the naked eye.

189. The Cone Nebula in Monoceros. Photographed in red light with the Palomar 200-inch reflector.

190. M.42, the Orion Nebula. This is a region in which fresh stars are being formed. Deep inside the Nebula is the infra-red source BN (the Becklin-Neugebauer Object) now believed to be a very young, powerful star whose visible light cannot penetrate the dust in the nebula. Reproduced by kind permission of the Karl Schwarzschild Observatory, Tautenburg, Germany.

well away from the centre of the system. We will say more about this in a few moments.

Messier catalogued a number of star clusters, both open and globular, but his list also included *nebulae*, which are clouds of dust and gas where fresh stars are being born. The most famous of all nebulae is M.42, in Orion's Sword, which is visible to the naked eye as a misty patch and is very obvious in binoculars. Telescopes will show bright and dark patches, and also the four hot stars making up the so-called Trapezium (Theta Orionis), which are illuminating the nebula and making it shine – and

also, incidentally, 'exciting' the nebular material and making it emit a certain amount of light on its own account.

Most of the gas is hydrogen, which comes as no surprise. The density is almost incredibly low. If you took a core sample right through the Nebula, which is over 100 light-years in diameter, you would not collect enough material to balance the weight of a pound coin.

Inside the Nebula is an object which sends out a great deal of infra-red radiation; it is known as BN, after its two discoverers, Eric Becklin and Gerry Neugebauer. It is an immensely luminous star, but we can never see it; its visible light is blocked out by the dust in the Nebula, and it will live and die unseen by us. There are other infra-red sources, too, as well as many young, irregularly variable stars which have only just started to shine, and have not evolved as far as the Main Sequence.

There are many other bright nebulae, some 'emission' nebulae of the same type as M.42, and others which are 'reflection nebulae', illuminated by stars not hot enough to make the material glow on its own account. If there are no suitable stars to light up the material, the nebula remains dark, and we can detect it only because it blots out the light of stars beyond. The most famous dark nebula is the Coal Sack in the Southern Cross. It can never be seen from Britain, but there are dark patches here and there in the Milky Way, particularly in Cygnus.

Messier's list included a few objects which are neither clusters nor ordinary nebulae. There are, for example, the *planetary nebulae*, which are simply old stars which have thrown off their outer layers, and look rather like small, dimly-shining cycle tyres. The best-known example is M.57, the Ring Nebula in Lyra, which is visible with a small telescope. The central star is very small and hot, but not easy to see except with large apertures. And there is one *super-nova remnant* – M.1, the Crab Nebula,

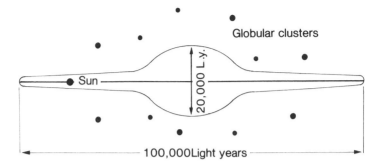

191. Shape of the Galaxy.

Sun

20,000 L.-y.

Globular clusters

100,000 Light years

which contains a pulsar, and sends out radiations at almost all wavelengths, from long radio waves to the ultra-short gamma-rays. It is 6000 light-years away.

Now let us come back to the Galaxy itself (Fig. 191).

It is, as we have seen, a flattened system, about 100,000 light-years in diameter, with a central bulge. When we look along the main plane, we see many stars in almost the same direction, and it is this which causes the Milky Way effect. The Sun's distance from the centre is a little less than 30,000 light-years. Unfortunately we cannot see all the way through to the centre, because there is too much interstellar dust in the way, but we know where it is; it lies beyond the magnificent star-clouds in Sagittarius.

What lies in the central region? We have to admit that we are uncertain. Radio waves and infra-red can be detected, and there is a distinct chance that close to the centre there is a massive black hole, but we cannot yet be sure.

From 'above' or 'below' the Galaxy would be seen to be spiral in shape, with the Sun lying near the edge of one of the spiral arms. This was first proved by radio observations. Clouds of cold hydrogen send out radiations at a wavelength of 21 cm; by plotting these clouds, we can trace the spiral arms. Not unexpectedly, the whole Galaxy is rotating. In the region of the Sun, the rotation period is of the order of 225,000,000 years.

Less than a century ago, it was still believed that our Galaxy was the only one. We know better now – and this leads us on to a consideration of the greater universe.

Questions

1. (a) What is the difference between an open cluster and a moving cluster?
   (b) Is the Pleiades cluster made up mainly of Population I or Population II? How do you know?
   (c) Name any three open clusters.
2. (a) What exactly is a globular cluster?
   (b) How were the distances of the globular clusters first measured?
   (c) The globular clusters are not distributed evenly all over the sky. Why not?
3. (a) What is the difference between an emission nebula and a reflection nebula?
   (b) A planetary nebula is neither a planet nor a true nebula. What, then, is it?
   (c) Why does the Coal Sack appear as a dark mass?
4. (a) Describe the position of the Sun in the Galaxy.
   (b) Explain the cause of the Milky Way effect.
   (c) Why cannot we see all the way through to the centre of the Galaxy?

Practical work

Here again there are many examples within the range of binoculars or small telescopes, and we have given only a very short 'short-list'.

1. *The Pleiades*. Fine open cluster. Use the naked eye to see how many stars you can count, and map them; then change to binoculars and note the appearance.

2. *The Hyades*. Rather scattered cluster; use the naked eye or binoculars. A V-formation of stars extending from Aldebaran.

3. *Praesepe*. Fine open cluster, well seen with binoculars and not difficult with the naked eye on a dark night. Its position is shown by the two stars Delta and Gamma Cancri.

4. *The Sword Handle in Perseus*. Double cluster; just visible with the naked eye, but to see it properly you will need binoculars or (preferably) a low-power magnification on a telescope.

5. *The Wild Duck Cluster in Scutum. M.11*. Fan-shaped open cluster, on the boundaries of Aquila and Scutum (the Shield). It is visible with the naked eye, but may be difficult to identify because of the richness of the Milky Way. Use a low-power eyepiece on a telescope.

6. *M.13, the Hercules Globular* (Fig. 192). Between Zeta and Eta Herculis, closer to Eta. It is difficult with the naked eye; binoculars show it clearly, and a small telescope will resolve its outer parts into stars.

7. *M.42, the Orion Nebula*. This has been described in the text. If you have a telescope of, say, 3-inch (7.6-cm) aperture you will see all four components of the illuminating star, the 'Trapezium' (Theta Orionis).

8. *The Ring Nebula in Lyra, M.57*. Planetary nebula, easy to find because it lies midway between the naked-eye stars Beta and Gamma Lyrae, near Vega. You will need a telescope to see it, and the central hot star requires at least 6-in in aperture. Some people claim to be able to see M.57 with binoculars, but I admit that I have never been able to do so.

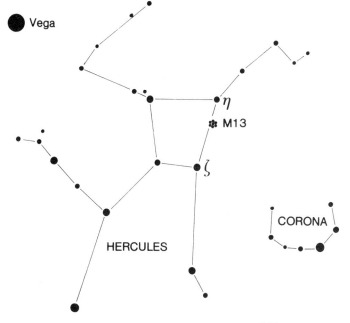

192. Finding M.13, the Globular Cluster in Hercules. Look between Corona and Vega; you will see the rather faint outlines of Hercules. Zeta is of magnitude 3. The cluster lies between Zeta and the fainter star Eta, rather closer to Eta. Unless the sky is very clear, you will not see it with the naked eye, but binoculars will show it easily.

23
Beyond the Milky Way

Look in Andromeda, marked by the line of stars leading off from the Square of Pegasus, and you may be able to make out a dim, misty patch. If you have binoculars you can find it easily enough. Telescopically, it appears as a somewhat elongated smudge of light. Photographs taken with large telescopes show it to be a spiral system, though it lies at an unfavourable angle to us and the main beauty of the spiral is lost. It was No. 31 in Messier's catalogue; we know it as the Andromeda Spiral – the only external galaxy clearly visible with the naked eye from Britain.

Messier's so-called nebulae were of two distinct kinds. Some looked like (and were) masses of dust and gas, such as M.42 in Orion's Sword, but others looked as though they were made up of stars. Could these 'starry nebulae' be galaxies in their own right, far beyond the Milky Way?

It was not easy to find out, because the objects were much too remote to show any parallax shifts. It was only in 1923 that the question was answered. Using the Mount Wilson 100-inch reflector, then much the most powerful telescope in the world, Edwin Hubble detected Cepheid variables in some of the starry nebulae, including M.31. As soon as he measured their periods and worked out their distances, he realised that the Cepheids – and, hence, the spirals in which they lay – could not possibly be members of our Galaxy. He gave the distance of M.31 as 750,000 light-years.

Actually, even this proved to be an under-estimate, because Hubble did not know that there are two types of Cepheids:

those of Population I, and those of Population II. He had taken them to be Population II, but in 1952 Walter Baade found that they were of Population I, which are twice as powerful – and, therefore, twice as far away. The modern value for the distance of the Andromeda Spiral is 2,200,000 light-years.

It is not the closest of all the galaxies. In the south of the sky, never visible from Europe, are the two Clouds of Magellan, which are conspicuous naked-eye objects and look rather like broken-off parts of the Milky Way, Both are within 200,000 light-years, and seem to be satellites of our Galaxy, though they are much smaller. Also within a few million light years are more than two dozen other systems, making up what we call the Local Group. There are three major spirals: in order of size, the Andromeda system (1½ times as large as our Galaxy), our Galaxy, and the Triangulum Spiral, which you can see with binoculars. The others are much smaller, and usually irregular. (One other large system, Maffei 1, may also be in the Local Group, but we know little about it, because it is almost hidden by dust lying in the plane of the Milky Way.)

Beyond the Local Group we come to vast numbers of other galaxies. Some are spiral, some elliptical, some irregular; some are larger than ours, others smaller. They contain objects of all kinds, ranging from double stars, variables, novae and open clusters to globular clusters and gaseous nebulae. There is absolutely no reason to suppose that our system is in any way exceptional.

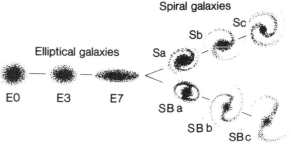

193. Hubble's classification of galaxies: ellipticals (E0 to E7), spirals (Sa, Sb, Sc) and barred spirals (SBa, SBb, SBc).

Hubble worked out a system of classification (Fig. 193). Spirals are designed Sa, Sb and Sc according to the 'tightness' of their arms; there are the curious barred spirals (SBa, SBb, SBc) in which the arms seem to come from the ends of a bar through the main plane; and there are

194. M.51 (NGC 5194), the Whirlpool Galaxy in Canes Venatici, with its satellite NGC 5195. Photographed from the Hale Observatories (200-inch reflector). This is the finest example of an open spiral, and was also the first spiral to be identified as such – by Lord Rosse in 1845. Its distance is 37,000,000 light-years.

ellipticals (E0 to E7). It used to be thought that this diagram indicated a sequence of evolution, so that a spiral might turn into an elliptical or vice versa, but this is wrong; everything depends upon the initial mass and speed of rotation of the galaxy.

The arms of a spiral are probably not permanent features. What seems to happen is that a 'pressure wave' sweeps round the system, and in this 'wave' the density of the material is above average, so that star formation is triggered off, and we can trace the arms themselves. As the 'wave' sweeps on, there are new regions of star formation. Needless to say, this takes a very long time indeed!

Even the Local Group is not unusual. Galaxies do tend to cluster; more than 50,000,000 light-years away we have the Virgo cluster, containing thousands of members, beside which the Local Group looks very small indeed. There are even 'superclusters' – clusters of clusters of galaxies.

Out to distances of several tens of millions of light-years we can use the Cepheids to measure distances, but with more remote systems the Cepheids are lost in the general background glow. We can go further by using supergiant stars, on the very reasonable assumption that supergiants in outer systems are likely to be about as powerful as supergiants in the Milky Way. There are also supernovae, which can be seen over even greater distances. (One supernova has been seen in M.31 – in 1885 – though unfortunately we have no reliable knowledge of its spectrum, because nobody at the time realised what it was; and more recently in 1987, a supernova flared up in the Large Cloud of Magellan, at 169,000 light-years, becoming a prominent naked-eye object for a few months.) But eventually we have to rely upon a much less direct method, involving the Doppler effect.

Early on – even before Hubble's great discovery in 1923 – it had been found that apart from the members of the Local Group, all the galaxies showed red shifts in their spectra. Since the spectrum of a galaxy is made up of the combined spectra of vast numbers of stars, they are bound to be something of a jumble, but the main lines are easy to measure, and the red shifts were very marked. Now, as we have seen, a red shift means a velocity of recession. The galaxies were moving away from us, and away from each other. The entire universe was expanding.

Hubble also found that there was a definite relationship between the distance of a galaxy and its velocity of recession. The formula is:

$$\frac{\Delta\lambda}{\lambda} = \frac{v}{c}$$

λ = the normal wavelength of the spectral line,
$\Delta\lambda$ = the observed shift in wavelength,
c = the velocity of light (300,000 km/sec or 3×10^5 km/sec^{-1}),
v = the required velocity.

Now, suppose that a line in the spectrum of a galaxy is red-shifted by 60 Ångströms from its normal value of 6000 Ångströms. How fast is the galaxy moving away?

Substituting in the formula,

$$\frac{60}{6000} = \frac{v}{3 \times 10^5}$$

and simple mathematics gives us a velocity of 300 kilometres per second.

If the shift had been to the blue, we could have made a calculation of the same sort, but no galaxy beyond the Local Group has a blue shift. It is true that at the present moment the Sun is actually approaching the Andromeda Spiral, but this is due entirely to our own motion round the centre of the Galaxy, and does not mean that M.31 and our Galaxy are approaching each other.

(Obviously, all our measurements of distances in the further reaches of the universe depend on the assumption that the red shifts are pure Doppler effects. If not, then our whole scale breaks down. There are some astronomers, notably Dr Halton Arp and Sir Fred Hoyle, who believe this to be the case. If they are right, then we will have to start again; but for the moment we are following the 'official' view.)

The so-called Hubble constant has an uncertain value, but is generally taken to be about 70 km per second per megaparsec – one megaparsec being one million parsecs. And you can probably see a problem arising. If the rule of 'the further, the faster' holds good, then we will eventually come to a distance at which a system is moving away from us at the full velocity of light. In this case we will be unable to see it, and we will have come to the boundary of the observable universe, though not necessarily the boundary of the universe itself. We cannot yet reach out as far as that, but we are coming within range of it, due largely to the power of radio methods.

Among the galaxies there are some which are very powerful in the radio range, though in most cases the main sources come from 'lobes' to either side of the visible galaxy (Fig. 196). The so-called Seyfert galaxies, which have large centres and weak spiral arms, are usually radio sources, and appear to have very active cores. But there are also the strange objects called *quasars*, which were originally identified because of their radio waves.

195. Remote galaxies. There are, of course, many foreground stars in our own system.

The quasar story began in 1963, when it was found that radio emissions were coming from objects which looked like faint blue stars. When the optical spectra of the objects were examined it was found that they were not stars at all; they were much more dramatic. Their spectra were un-starlike, and all the lines were tremendously red-shifted. This meant that they must be very remote – and yet they seemed almost stellar in appearance.

As the measurements improved, the mystery deepened. The objects – originally called 'quasi-stellar sources' – were in most cases thousands of millions of light-years away, and were receding at high velocities. As time went by it was found that by no means all quasars are strong radio emitters, but all were super-luminous, shining as brilliantly as thousands of normal galaxies combined.

Quasars are certainly the most distant objects known. The individual record holder seems to change annually, but many are now known to be thousands of millions of light years away from us and appear to be receding at well over 90 per cent of the speed of light. More and more observational evidence is mounting which supports the theory that these objects are merely the centre regions of extremely active galaxies, almost certainly powered by black holes. The conclusion is not yet certain, but we have detected the outer regions of the galaxy in a few cases and have no reason to suspect them to be missing in the others. The mystery of why the galaxies are so active remains open, of course.

All of this brings us about as far from our home galaxy as we can go. We now need to step back and look at the universe as a whole; and that is the task of cosmology.

196. Radio 'lobes' to either side of a visible galaxy.

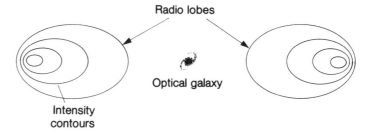

Radio lobes

Optical galaxy

Intensity contours

185

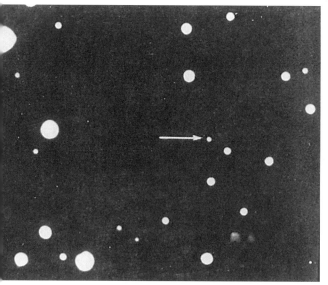

197. The quasar OQ 471 – well over 10,000 million light-years away. The quasar appears as a dot at the end of the arrow indicator.

198. Barred galaxy NGC 1365, imaged by the Very Large Telescope in Northern Chile.

Questions

1. (a) How did Edwin Hubble measure the distances of galaxies?
   (b) Why was his value for the distance of the Andromeda Spiral much too low?
   (c) Name any three members of the Local Group of galaxies.
2. (a) 'All galaxies are receding from the Milky Way galaxy.' Is this a true statement?
   (b) Draw a diagram to illustrate Hubble's system of classifying galaxies.
   (c) Draw a diagram to illustrate the areas of radio emission in a radio galaxy.
3. What is the recessional velocity of a galaxy, if the wavelength of a spectral line at 8000 Å is increased by 200 Å?
4. What are the main characteristics of quasars?
5. How will we eventually be able to tell whether or not the present expansion of the universe will continue indefinitely?

Practical work

Without a telescope of reasonably large aperture, and considerable specialist knowledge, there is little practical work that can be carried out with regard to galaxies. However, you may look for and identify the Andromeda Galaxy, M.31, with the help of Fig. 199. It has two much smaller companion galaxies, M.32 and NGC 205, both of which are elliptical. They are not, however, visible with binoculars of the 7 x 50 variety.

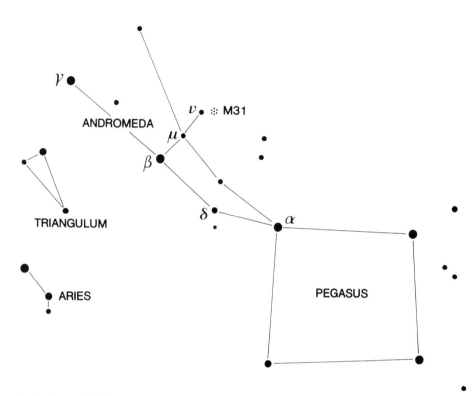

199. Finding the Andromeda Galaxy.

24
Cosmology

If astronomy is the study of everything outside the Earth's atmosphere, then cosmology is the study of the universe considered as a whole. In other words, instead of confining ourselves to the structure of individual galaxies, for example, or the processes of star formation, we are now considering the history of the universe as a whole; its birth, its evolution since and eventually its possible end.

The old view, which prevailed from ancient times right up until the last century, was of a universe that was, on a large scale, static, unchanging and infinite. It must be said that although this view was held primarily on religious grounds, there were no obvious scientific contradictions either. One of the first astronomers to think seriously about alternatives was Heinrich Olbers, who we remember today as author of the famous Olbers' Paradox. Essentially, he pointed out that if the universe is infinite then, no matter in which direction we chose to look, we would eventually see a galaxy. Therefore the night sky should be bright, which it is most definitely not! (In fact, the paradox can be stated in a more technical way to avoid the objection of light being absorbed over the great distances involved, but there is no need to go into such detail at GCSE.)

It was thus clear that there was a problem with our view of the universe, but the solution remained obscure until the early years of the twentieth century. Detailed observations of the spectra of galaxies by Edwin Hubble revealed that the spectral lines of all but the closest were *redshifted*, indicating that the entire universe appeared to be rushing away from us. (Note, however, that the very closest galaxies – those in our Local Group – are not necessarily receding, and in fact the Andromeda Galaxy, for example, is approaching us at the present time.) In fact, the same effect would be seen anywhere in the universe, and so we reach the view that space itself is expanding. (Do not be misled into thinking that somewhere there must be a centre of the expansion. The situation is rather analogous to blowing up a spotty balloon. The spots on the surface of the balloon rush away from each other, but no one spot is in the centre.) That, of course, means that there must have been a time in which the expansion began; the famous 'Big Bang'.

Despite a few dissenting voices, the evidence for the big bang now looks to be fairly conclusive. Aside from the expansion itself, astronomers have been able to successfully predict from the theory the relative abundances of hydrogen and deuterium (deuterium has the same chemical properties as hydrogen but contains an extra neutron in its nucleus) that we observe in the universe today. The clinching evidence, however, was the discovery of the cosmic microwave background radiation. As the universe expanded, the energy left over from the big bang cooled until the present day, where we observe it as (almost) uniform background radiation corresponding to a temperature of 2.7K (2.7 degrees above absolute zero). Crucially, the satellite

COBE was, in 1992, able to detect slight variations in the radiation, and these are believed to be the seeds which eventually became the clusters of galaxies that populate the present day universe. A further mission, Planck, is planned for 2007 to give us a much more detailed look at the microwave background.

An obvious question is the age of the universe; in other words, how long ago did the big bang actually occur? Astronomers have been trying to answer this almost since the publication of the initial theory, and it all rests on the value of something known as the Hubble constant. Hubble discovered that the further away a galaxy was, the faster it was receding from us. This means that for any galaxy, the value of the distance divided by the speed of recession should always be the same, and this is the Hubble constant. Recent observations, appropriately enough with the Hubble Space Telescope, seem to be converging on a value of around $70 \text{kmsec}^{-1}\text{Mpc}^{-1}$, which corresponds to a universe between 12,000 and 15,000 million years old.

So much for the past; what of the future? Here the crucial measurement to be made is the total mass of all the matter in the universe. We know that at present the universe is expanding outwards and the only thing that can halt the expansion is the force of gravity. If the total mass of the universe is less than the critical value (usually expressed as $\Omega < 1$ where Ω represents the mass of the matter in some convenient units and 1 is the critical value), then gravity will not be able to halt the expansion and the universe will continue expanding forever. If the total mass lies exactly on the critical value ($\Omega = 1$) then the expansion will simply stop, whereas if the total mass exceeds the critical value ($\Omega > 1$) then the universe will eventually collapse in a 'Big Crunch'. So far, we are certain only that the visible matter in the universe is far less than that needed to halt the expansion. However, there does appear to be (from other evidence) a great deal of 'dark matter' out there which may or may not make up the shortfall. Our best estimate so far is that Ω is close to 1, and so the fate of the universe is literally hanging in the balance. Further observations over the next few years should do much to pin down the value far more accurately.

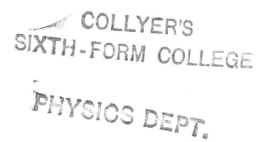

25

Life Elsewhere?

Of all the questions put to us, one of the commonest is: 'Can life exist beyond the Earth? If so, what will it be like?'

First, we must decide just what we mean by 'life'. We know a great deal about living matter, and we know that only one type of atom – carbon – can join with other atoms to make up the complicated molecules necessary for life. If we are wrong about this, then most of our modern science is wrong too. As soon as we start to discuss entirely alien life-forms, of the type known to science-fiction writers as BEMs (=Bug-Eyed Monsters), speculation is not only endless, but also pointless. It is far more sensible to limit ourselves to life of the sort we can understand.

This being so, we can rule out almost all the planets in our Solar System apart from the Earth, because they simply do not meet the requirements. Life cannot exist without the right sort of atmosphere, and this leaves only Mars and Titan. The Martian atmosphere is painfully thin, and in any case is mainly carbon dioxide; if any life survives there, it must be very lowly. Titan is presumably ruled out by its very low temperature and the fact that its atmosphere contains a great deal of methane.

Yet we know that the Sun is only one of a hundred thousand million stars in our Galaxy alone, and there are millions upon millions of other galaxies. It is surely absurd to suppose that in all this host, only our undistinguished Sun is attended by an inhabited planet. As yet we cannot definitely prove that other planetary systems exist, but the evidence is starting to build up; IRAS, the Infra-Red Astronomical Satellite which operated for most of 1983, found that many nearby stars are associated with cool material which could be planet-forming. Among these stars are Vega and Fomalhaut, while in one case, that of the southern star Beta Pictoris, the material has actually been photographed optically.

A massive planet orbiting a relatively low-mass star should produce measurable perturbations in the star's movement, and effects of this kind have recently been found in several fairly nearby stars. The first detection of a planet orbiting another star was made in 1995 by two Swiss astronomers, M. Mayor and D. Queloz; the star concerned was 51 Pegasi, which is 54 light-years away. The mass of the planet was found to be about half that of Jupiter, and by now there are more than a dozen similar cases. No extra-solar planets have actually been seen, and it cannot yet be said that we have final proof of their existence, but the evidence is very strong indeed.

We must then ask: what are the prospects of life elsewhere? Of course, we cannot claim that life will appear wherever conditions are suitable for it – but why not? If we have a planet like the Earth, orbiting a star like the Sun, it is surely not unreasonable to suggest that Earth-type life may be expected.

Unfortunately, there is no prospect of

interstellar travel by means of rockets. The time taken to reach even the nearest star would be hopelessly long. If flight to the stars is ever to be achieved, it must follow some fundamental 'breakthrough' which is quite beyond us at the moment. We may be as far from it as King Canute was from television.

Efforts have been made to detect rhythmical signals from other systems which might be interpreted as deliberate transmissions, but, not surprisingly, without success. And we can safely ignore the numerous flying saucer or UFO stories of recent years. A visit from alien beings is not impossible, but there is no evidence that it has happened yet.

*

This, for the moment, is as far as we can go. We have done our best to present you with a picture which will help you to pass the GCSE examination if you feel so inclined; but even if you do not want to take the examination, or even to make a serious hobby out of astronomy, we hope that you have been interested in what we have had to say.

We wish you the best of luck.

Appendix
Answers to Questions: Chapter 5

3. Here we will strive for accuracy, at least with regard to the difference between solar and sidereal time. Relative to the stars, the Earth makes one orbital circuit in 366 days. Relative to the Sun, this takes 365 days, due to the revolution of the Earth; therefore Sun time runs slow on star time, and the difference is 1 part in 365. In one hour, 1 part in 365

$$\frac{= 60 \times 60}{365} = 9.87 \text{ seconds}$$

Generally we can round this off to 10 seconds; if so, the increment for 18 hours in the following sum will be 3 minutes instead of 2m 57s.

		h	m	s
(a) The GST at 0 hours	=	12	54	00
Increment for 18 hours	=	2	57	
		12	56	57

		h	m	s
Therefore, the GST at 18 hours	=	12	56	57
		18	00	00
		30	56	57

This is actually on the following day (6h 56m 37s on 7 April) but we will simplify matters by leaving it as it is. We now allow for the longitude of Broadstairs, which is 1°26'E. Multiply by 4, and we obtain 5m 44s.

		h	m	s
GST at 18h	=	30	56	37
Long. of Broadstairs: allow			5	44
LST at Broadstairs	=	31	02	41

We must add, because Broadstairs is east of Greenwich.
Hour angle = LST – RA

		h	m	s
LST	=	31	02	41
RA at Betelgeux	=	5	53	00
Hour angle	=	25	09	41

We must now write this conventionally, so it becomes 1h 9m 41s on 7 April.

(b) Taking the latitude of Broadstairs away from 90°, we obtain the co-latitude, 38°39'. To this we add the declination of Betelgeux, because Betelgeux is north of the celestial equator. 38°39' + 7°24' = 46°03', which is therefore the meridian altitude of Betelgeux.

(c) To be circumpolar, the co-declination of Vega (i.e. its declination subtracted from 90°) must be less than the observer's latitude. The co-declination of Vega is 51°16', and the latitude of Broadstairs is 51°21'. Therefore, Vega is just circumpolar.

4. (a) We need the co-latitude of Bristol, i.e. its latitude subtracted from 90°. 90° − 51°27' = 38°33'. To this we must add the declination of Altair, which is north of the equator. 38°33' + 8°44' = 47°17', which is therefore the meridian altitude of Altair.

(b) The LST = the RA of the star = 19h 48m.

		h	m	s
LST of transit	=	19	48	00
Allow for long. of Bristol	=		10	12
GST of transit	=	19	58	12

(c) The difference is 3m 56s per day, so in 7 days this amounts to 27m 32s. Taking this from 22h 20m, we obtain 21h 52m 28s, which will be the transit of Altair one week later.

5. (a) We know the LST of the transit, since it must be equal to the RA of Mizar: 13h 21m 54s. Adding to allow for the longitude of McGill (i.e. dividing this longitude by 15) we obtain the GST of the transit: 18h 16m 13s. Working in approximations, we simply subtract 10h 35m 26s, which was the GST at 0 hours, and obtain a value of 7h 40m 47s, which will be the GMT of the transit at 0 hours at McGill. More accurately, we must add on 1m 16s to the GST at 0 hours to allow for the difference between sun and star time. This gives us 10h 36m 42s, and subtracting this from the GST of the transit of Mizar we obtain an answer of 7h 39m 31s, which is correct to the nearest second.

(b) We need the co-latitude of McGill. 90° − 45°30'20" = 44°29'40".

		h	m	s
Co-latitude of McGill	=	44	29	40
Add N. dec. of Mizar	=	55	11	00
Altitude of Mizar	=	99	40	40

But this exceeds 90°; Mizar is 'over the top' by 9°40'40", and this must be subtracted from 90°, giving us an answer for the actual altitude of Mizar = 80°19'20".

(c) We know that the GMT is 20h 0m. To obtain the local mean time we must allow for McGill's longitude, subtracted because McGill is west of Greenwich. As we have found in part (a), the allowance is 4h 54m 19s.

		h	m	s
GMT	=	20	00	00
Longitude allowanc	=	4	54	19
Local mean time	=	15	05	41

The GST at 0 hours has been given as 10h 35m 26s. The increment (sun time compared with star time) for 20 hours is 3m 17s.

		h	m	s
Local mean time	=	15	05	41
GST at 0h	=	10	35	26
		25	41	07
Increment	=		3	17
LST	=	25	44	24

To obtain the hour angle of Mizar, we simply subtract the star's RA:

		h	m	s
LST	=	25	44	24
RA of Mizar	=	13	21	54
LHA of Mizar	=	12	22	30

which is the local hour angle of Mizar as seen from McGill Observatory at 20h GMT. If less accuracy were needed, the longitude allowance could be rounded off to 4h 54m instead of 4h 54m 19s.

Brief Bibliography

Arnold, H.J.P., *Night Sky Photography* (Philip).

Arnold, H.J.P., Doherty, P., and Moore, P., *Photographic Atlas of the Stars* (Institute of Physics Publishing).

Clark, S., *Toward the Edge of the Universe* (John Wiley).

Fischer, D., and Dubreck, H., *Hubble* (Copernicus Press).

Kaufmann, W., *Universe* (Freeman).

Lovell, B., *Out of the Zenith* (Cambridge).

Malin, D., and Murdin, P., *Colours of the Stars* (Cambridge).

——— *Colours of the Galaxies* (Cambridge).

Moore, P., *Atlas of the Universe* (Philip).

——— *Exploring the Night Sky with Binoculars* (Cambridge).

——— *Patrick Moore on Mars* (Cassell).

——— *The Observer's Year* (Springer).

Mobberley, M., *Astronomical Equipment for Amateurs* (Springer).

Nicolson, I., *The Universe Unfolded* (Cambridge).

Yeomans, D., *Comets* (John Wiley).

various authors, *Small Astronomical Observatories* (Springer).

Spry, R., *Make Your Own Telescope* (South Downs Planetarium, Chichester).

The *Yearbook of Astronomy* is published annually by Macmillan.

Periodicals include the monthly *Sky and Telescope* (American, but readily available in Britain) and *Astronomy and Space* (Irish, also readily available in England).

All enthusiasts are recommended to join the British Astronomical Association which publishes a regular Journal and an annual Handbook. Its address is Burlington House, Piccadilly, London W1.

Glossary

absolute magnitude. The apparent magnitude that a star would have if it were seen from a standard distance of 10 parsecs, or 32.6 light-years.

achromatic object-glass. An object-glass corrected for false colour; it is made up of two separate components, made of different types of glass.

albedo. The reflecting power of a body, from 0 (black) to 100 (a perfect reflector).

altazimuth mount. A type of telescope mount in which the telescope can be moved freely in both altitude and *azimuth*.

altitude. The angular distance of a celestial body above the horizon, from 0° at the horizon to 90° at the zenith.

Ångström unit. One hundred-millionth part of a centimetre.

aphelion. The position of a planet or other body when at its greatest distance from the Sun.

asteroids (minor planets). Small planets, most of which move round the Sun between the orbits of Mars and Jupiter.

astronomical unit. The mean distance between the Earth and the Sun: 149,598,500 km or, in round numbers, 150,000,000 km.

aurorae. Glows in the upper atmosphere, due to charged particles from the Sun: Aurora Borealis in the northern hemisphere, Aurora Australis in the southern.

azimuth. The angular bearing of a celestial object, measured from north (0°) through east (90°), south (180°) and west (270°) back to north (360° or 0°).

barycentre. The centre of gravity of the Earth-Moon system. It lies within the Earth's globe.

big bang theory. The theory that all the matter in the universe came into existence at the same moment, between 15,000 million and 20,000 million years ago; 'space' was created at the same time.

binary star. A star made up of two components, physically associated.

black dwarf. A dead star, which has used up all its energy. It is not certain whether the universe is yet old enough for any black dwarfs to have been formed.

black hole. A region round an old, massive, collapsed star from which nothing – not even light – can escape.

Bode's law. A mathematical series linking the distances of the planets from the Sun. It has probably no real significance.

captured rotation (or *synchronous rotation*). If the axial rotation of a body is equal to its revolution period round its primary, the rotation is 'captured', and the body keeps the same face permanently toward its primary.

Cassegrain reflector. A type of reflecting telescope in which the light from the object under observation is sent from the main mirror on to a convex secondary mirror, and is then reflected back to the eyepiece through a hole in the main mirror.

celestial sphere. An imaginary sphere surrounding the Earth, whose centre is the same as that of the Earth.

Cepheid. A short-period variable star. The variations are perfectly regular; the period is linked with the real luminosity of the star.

charge-coupled device (CCD). A very sensitive electronic device, far more sensitive than a photographic plate.

chromosphere. That part of the Sun's atmosphere lying above the bright surface or photosphere. It consists mainly of hydrogen.

circumpolar star. A star which never sets from the place of observation.

clusters, stellar. Groups of stars which have a common origin.

colour index. A measure of a star's colour; the difference between the photographic magnitude and the visual magnitude – positive if the star is reddish, negative if the star is bluish, zero if the star is white.

colures. Great circles on the celestial sphere. The equinoctial colure passes through right ascension 0 hours and 12 hours; the solsticial colure passes through right ascension 6 hours and 18 hours.

conjunction. (a) A planet is in conjunction with a star, or another planet, when it passes close by it in the sky; it is of course a line-of-sight effect. (b) A planet is in superior conjunction when it is on the far side of the Sun with respect to the Earth, and in inferior conjunction when passing between the Sun and the Earth. Obviously, only Mercury and Venus can pass through inferior conjunction.

constellation. A group of stars named after a mythological character or an animate or inanimate object. Since the stars in a constellation are at very different distances from us, the pattern is of no real significance.

Copernican System. The system in which the Sun is the central body, with the planets moving round it. Also known as the **heliocentric system** (Greek, *helios*, 'sun').

corona. The outermost part of the Sun's atmosphere, made up of thin gas at a very high temperature.

coronal hole. A region of exceptionally low density in the solar corona. Charged particles escape through these holes and produce the solar wind.

cosmic rays. These are not rays at all, but high-speed atomic particles from outer space. The heavy cosmic rays are broken up in the upper air, and only their fragments reach the ground.

cosmic year. The time taken for the Sun to complete one orbit round the centre of the Galaxy: about 225,000,000 years.

Coudé system. An optical system in which the light from the body under observation is received in a fixed direction.

counterglow. The English name for the faint sky-glow known more commonly as the Gegenschein.

culmination. The maximum altitude of a celestial body above the horizon.

day. The time taken for the Earth to spin once on its axis, with respect to the stars (sidereal day), the Sun (solar day) or the mean sun (mean solar day).

declination. The angular distance of a celestial body north or south of the celestial equator.

degree of arc. A unit for measuring angles. A full circle contains 360°; each degree contains 60 minutes of arc, and each minute contains 60 seconds of arc.

density. The amount of matter in a unit volume of substance. For most cases water is taken as unity. The mean density of the Earth is 5.52 times that of water (this is the Earth's **specific gravity**).

dichotomy. The time of exact half-phase of the Moon, Mercury or Venus.

direct motion. Bodies which move round the Sun in the same sense as the Earth; the term is also used for satellites of other planets.

diurnal motion. The apparent daily rotation of the sky, due to the real rotation of the Earth.

Doppler effect. The apparent change of wavelength of light (or sound) caused by the motion of the source with respect to the observer. If the light-source is approaching, the wavelength is shortened (blue shift); if receding, the wavelength is increased (red shift).

double star. A star made up of two components. Double stars may be *optical*, due to line-of-sight effects, or *binary* systems, which are physically associated.

early type stars. Stars of spectral type W, O, B and A. The name was given when it was still thought that the spectral sequence was a true evolutionary sequence.

earthshine. The dim visibility of the non-sunlit side of the Moon, due to light reflected on to the Moon from the Earth.

eclipses, lunar. The entry of the full moon into the shadow cast by the Earth. They may be either total or partial.

eclipses, solar. The temporary blotting-out of the Sun when the new moon passes in front of it. Solar eclipses may be total, partial or annular.

eclipsing binary (or eclipsing variable). A binary system in which, as seen from the Earth, the components pass regularly in front of each other and cause the apparent magnitude of the system to drop.

ecliptic. The projection of the Earth's orbit on to the celestial sphere; it may also be defined as the apparent yearly path of the Sun against the stars. It is inclined to the celestial equator by 23½°.

electromagnetic spectrum. The full range of wavelengths, from the very long radio waves through to the ultra-short gamma-rays. Visible light makes up only a very small part of the electromagnetic spectrum.

electron. A fundamental particle carrying unit negative charge.

element. A substance which cannot be split up chemically into simpler substances.

elongation. The apparent angular distance of a planet or comet from the Sun. At opposition, elongation is 180°. For the inferior planets this can never occur; the maximum elongation for Mercury is 28°, for Venus 47°.

emission spectrum. A spectrum consisting of isolated bright lines, each of which is characteristic of a particular element or group of elements.

ephemeris. A table showing the predicted positions of a moving celestial body.

epoch. A date chosen for reference purposes in quoting astronomical data.

equation of time. The interval by which the real Sun is ahead of or behind the mean sun. It can never exceed 17 minutes.

equator, celestial. The projection of the Earth's equator on to the celestial sphere. It divides the sky into two hemispheres.

equatorial mounting. A mounting in which the telescope is set upon an axis which is parallel to the axis of the Earth. When moved in azimuth, the altitude looks after itself.

equinox. A point where the ecliptic cuts the celestial equator. The Sun reaches the vernal equinox (First Point of Aries) about 21 March, and the autumnal equinox (First Point of Libra) about 22 September.

escape velocity. The minimum velocity at which an object must move in order to escape from the surface of a planet, or other body, without being given any extra impetus.

exosphere. The outermost part of the Earth's atmosphere.

extinction. The dimming of the light from a celestial body when near the horizon. It amounts to 3 magnitudes for a star 1° above the horizon, but 1 magnitude for a star at 10°, and above 45° it is very slight.

eyepiece (or **ocular**). The lens, or combination of lenses, at the eye-end of a telescope. It is responsible for magnifying the image produced by the object-glass or mirrors.

faculae. Bright, temporary patches above the Sun's bright surface or photosphere. They are usually associated with sunspot groups.

false colour technique. The use of colours on an image or chart in order to help in analysis.

finder. A small, wide-field telescope attached to a larger one, to help in locating target objects.

fireball. An exceptionally bright meteor (above magnitude −5).

First Point of Aries. The vernal equinox.

flares, solar. Brilliant outbreaks in the Sun's atmosphere, usually associated with active spot-groups. They emit charged particles and short-wave radiation.

focal length. The distance between the centre of a mirror (or lens) and the focus.

focus. The point where rays of light meet after having been converged by a lens or mirror.

Fraunhofer lines. The dark lines in the spectrum of the Sun.

free fall. The normal state of motion of an object in space under the influence of the pull of a central body. The earth is in free fall round the Sun.

galaxies. Independent systems of stars, far beyond our Milky Way system.

Galaxy, the. The system of stars of which our Sun is a member. It is often called the Milky Way Galaxy.

Galilean satellites. The four large satellites of Jupiter: Io, Europa, Ganymede and Callisto.

gamma-rays. Very short-wave electromagnetic radiations, with wavelengths of 10^{-12} metres or less.

gauss. The standard unit of measurement for a magnetic field. The magnetic field at the Earth's surface ranges between 0.3 and 0.6 gauss.

Gegenschein. A faint sky-glow, seen opposite to the Sun, and due to the illumination of thinly-spread interplanetary material.

geocentric theory. The old theory that the Earth was the central body of the Solar System.

gibbous phase. The phase of a body, shining by reflected light, when between half and full.

gnomon. The pointer of a sundial. The shadow of the gnomon on the dial gives the time. Also known as a **style**.

granules, solar. Features of the Sun's bright surface or photosphere. The granules last for only about 10 minutes each, and represent upcurrents.

gravitation. The force of attraction which exists between all particles of matter in the universe. If F is the attractive force between two bodies, m_1 and m_2 their masses, d is the distance between them, and G is the gravitational constant, then

$$F = G\frac{m_1 m_2}{d^2}$$

great circle. A circle on the surface of a sphere (such as the Earth) whose plane passes through the centre of the sphere.

Greenwich Mean Time (GMT). The local time at Greenwich, reckoned according to the mean sun.

Greenwich meridian. The line of longitude which passes through the Old Royal Observatory in Greenwich Park. It marks longitude 0°.

Gregorian calendar. The calendar now in use.

Gregorian telescope. An obsolete type of reflector in which the secondary mirror was concave, and the image was sent back to the eyepiece through a hole in the main mirror. Unlike the Cassegrain, it gave an erect image.

H.I and H.II regions. Clouds of hydrogen in the Galaxy. In H.I regions the hydrogen atoms are complete; in H.II regions they are ionised by the radiation from very hot stars. H.II regions shine as gaseous nebulae.

HR Diagram (or **Hertzsprung-Russell Diagram**). A diagram in which stars are plotted according to their spectral types and their absolute magnitudes. If colour index is used as one scale, it becomes a *colour-magnitude diagram*.

Halley's Comet. The only bright periodical comet. It has a period of 76 years, and is next due at perihelion in 2061.

halo, galactic. The spherical-shaped cloud of stars around the main Galaxy.

Harvest Moon. The full moon nearest to the autumnal equinox.

heliacal rising. The rising of a celestial body at the same time as sunrise. Generally taken to mean the date when the body first becomes visible with the naked eye in the dawn sky.

heliocentric theory. The theory according to which the Sun lies in the centre of the Solar System.

helium. The second lightest element. Its atom has two electrons.

horizon. The great circle on the celestial sphere which is everywhere 90° from the overhead point or zenith.

hour angle. The time which has passed since a celestial object crossed the meridian.

hour circle. A great circle on the celestial sphere passing through both celestial poles. The zero hour circle coincides with the observer's meridian.

Hubble constant. A constant relating to the recessional velocity of galaxies. It is usually taken as 55 kilometres per second per megaparsec.

Hubble time. The time which has elapsed since the origin of the universe.

Hunter's Moon. The full moon following Harvest Moon.

immersion. The entry of a celestial object into occultation or eclipse.

inferior conjunction. The position of an inferior planet (Mercury or Venus) when between the Earth and the Sun. Some comets and asteroids can also pass through inferior conjunction.

infra-red radiations. Electromagnetic radiations with wavelength between 10^{-6} and 10^{-4} metres.

ion. An atom which has lost one or more of its planetary electrons. The process of producing an ion is termed *ionisation*.

ion tail. The straight, gaseous tail of a comet.

ionosphere. The region of the Earth's atmosphere above the stratosphere. It contains the layers which reflect some radio waves back to the ground, making long-range communication possible.

IRAS. The infra-Red Astronomical Satellite. It operated for some months during 1983.

isophote. A line on a diagram joining all points of equal intensity or density.

Julian day. A count of the days, reckoning from 12 noon on 1 January 4713 BC. (The name has nothing to do with Julius Caesar!)

Kepler's Laws. Three important Laws of Planetary Motion, announced by Johannes Kepler between 1609 and 1618. They are described in the text.

kiloparsec. One thousand parsecs (3,260 light-years).

late-type stars. Conventionally, stars of spectral types M, R, N and S.

librations. The apparent tilts which enable us to see, at one time or another, 59 per cent of the Moon's surface from the Earth. They are described in the text.

light-curve. A graph showing the changing brightness of a variable star, or other body which changes in brilliancy.

light-year. The distance travelled by light in one year: 9.4607 million million kilometres.

local group. The group of galaxies of which our Galaxy is a member. Systems in the Local Group are the only galaxies not receding from ours.

lunation. The interval between successive new moons: 29d 12h 44m.

magnetic storm. A sudden disturbance of the Earth's magnetic field, due to charged particles sent out by the Sun.

magnetosphere. The area round a celestial body in which the magnetic field of that body is dominant.

magnitude. (a) Apparent: the apparent brightness of a celestial body – the lower the magnitude, the brighter the object. (b) Absolute: the apparent magnitude which an object would have if seen from a distance of 10 parsecs. (c) Photographic: the magnitude measured from the apparent size of the body on a photographic plate.

Main Sequence. The belt on the HR Diagram running from upper left to lower right. Most stars lie on the Main Sequence.

Maksutiv telescope. An astronomical telescope which makes use of both mirrors and lenses.

Maunder minimum. The period between 1645 and 1715, when sunspots were rare.

mean sun. An imaginary body travelling eastward along the celestial equator, at a rate of motion equal to the average rate of the real Sun along the ecliptic.

meridian, celestial. The great circle on the celestial sphere which passes through the zenith and both celestial poles. It cuts the observer's horizon at the exact north and south points.

Messier numbers. The numbers allotted to clusters and nebulae in the catalogue drawn up by Charles Messier in 1781.

meteor. A small particle, usually smaller than a grain of sand, which is seen when it enters the Earth's upper air and is burned away by friction. Meteors are cometary débris.

meteorite. A solid body which strikes the Earth from space. Most are iron, stone or a mixture of both materials. They are not associated with comets, and probably come from the asteroid belt.

micrometeorites. Extremely small particles, not massive enough to produce luminous effects when they enter the upper atmosphere.

micron. A unit of length: one-thousandth of a millimetre. There are 10,000 Ångströms to one micron. The usual symbol is the Greek letter μ.

microwaves. Electromagnetic radiations, intermediate between infra-red and radio waves; wavelengths 1 millimetre to 1 metre.

Milky Way. The luminous band crossing the sky, due to the fact that in this direction we are looking along the main plane of the Galaxy.

minor planets. Another (and more suitable!) name for asteroids.

minute of arc ('). One-sixtieth of a degree.

Mira stars. Variable stars with periods of a few weeks or months. They are named after the prototype star, Mira Ceti.

molecule. A stable association of atoms. Thus a water molecule consists of two hydrogen atoms with one atom of oxygen (H_2O).

month. (a) Calendar: the civil month – 30 or 31 days (28 or 29 in February). (b) Sidereal: the time taken for the Moon to complete one orbit: 27.32 days.

moving clusters. Groups of stars sharing a common motion through space.

multiple star. A star made up of more than two physically-associated components (e.g. Theta Orionis, Epsilon Lyrae).

NASA. The National Aeronautics and Space Administration (USA).

NGC. The New General Catalogue of Clusters and Nebulae, compiled by J.L.E. Dreyer a century ago.

nadir. The point on the celestial sphere immediately below the observer, and therefore opposite to the observer's zenith.

nanometre. One thousand millionth of a metre.

neap tide. The tide produced when the Sun and Moon are at right angles to the Earth.

nebula. A mass of 'dust' and tenuous gas in space.

Reflection nebulae shine only by reflected starlight; with emission nebulae, the associated stars are hot enough to make the material emit a certain amount of light on its own account.

neutrino. A fundamental particle with no electrical charge and, apparently, no rest mass.

neutron. A fundamental particle with no electrical charge, but a mass practically equal to that of a proton.

neutron star. A star made up mainly of neutrons – the remnant of a supernova outburst. Neutron stars rotate rapidly, sending out pulsed radio waves (**pulsars**).

Newtonian reflector. The most common type of reflecting telescope, in which the light is reflected from the main mirror on to a secondary flat mirror and thence to the eyepiece.

nodes. The points at which the orbit of the Moon, a planet, comet or asteroid cuts the plane of the ecliptic.

nova. A stellar outburst in the white dwarf component of a binary system.

object-glass (or **objective**). The main lens of a refracting telescope. Most objectives are compound, made up of two or more components.

obliquity of the ecliptic. The angle between the ecliptic and the celestial equator: 23°26'24", usually given as 23½°.

occultation. The covering-up of one celestial body by another. Strictly speaking, a solar eclipse is an occultation of the Sun by the Moon.

ocular. Alternative name for an eyepiece.

Olbers' paradox. A paradox discussed by H. Olbers in 1825: Why is it dark at night? It has been described in the text.

Oort cloud. A theory in which there is a cloud of comets orbiting the Sun at a distance of about a light-year.

open cluster. A loose galactic cluster of stars.

opposition. The position of a planet when exactly opposite to the Sun in the sky.

optical double. A double star in which the two components are not genuinely associated, but merely lie near the same line of sight.

optical window. The region of the electromagnetic spectrum in which radiations can pass through the atmosphere and reach the Earth's surface. It extends from about 300 to 900 nanometres (3,000 to 9,000 Ångströms).

orbit. The path of a celestial body.

orrery. A model showing the Solar System, with the planets capable of being moved round the Sun at their correct relative velocities.

parallax, trigonometrical. The apparent shift of an object when observed from two different directions.

parsec. The distance at which a star would show a parallax of 1 second of arc: 3.26 light-years, 206,265 astronomical units, or 30.857 million million kilometres.

penumbra. (a) The area of partial shadow to either side of the main cone of shadow cast by the Earth. (b) The lighter part of a sunspot.

perigee. The position of the Moon in its orbit when closest to the Earth.

perihelion. The position of a body in the Solar System when at its closest to the Sun.

perturbations. The disturbances in the orbit of a celestial body produced by the gravitational pulls of other bodies.

phase angle. The angle between the Earth and the Sun, as seen from another body in space.

phases. The apparent changes in shape of a celestial body from new to full.

photoelectric cell. An electronic device. Light falls upon the cell, and produces an electric current; the strength of the current depends upon the intensity of the light.

photoelectric photometer. An instrument used for measuring the brightness of celestial objects. It consists basically of a photoelectric cell used together with a telescope.

photon. The smallest 'unit' of light.

photosphere. The bright surface of the Sun.

plages, solar. Brighter regions on the Sun's surface, observed in the light of one element only (hydrogen or calcium). Also called *flocculi*.

planetary nebula. A small, hot dense star surrounded by a shell of gas.

plasma. Gas in which the atoms are wholly ionised.

Pogson's ratio. The ratio between the brightness of two stars at successive magnitudes. It is logarithmic; the ratio is 2.512 (since 2.512 is the fifth root of 100).

poles, celestial. The north and south points of the celestial sphere.

populations, stellar. Two main types of star regions: (I) in which the brightest stars are hot and white, (II) in which the brightest stars are old red giants and supergiants.

position angle. The apparent direction of one object with reference to another, from north (0°) through east (90°), south (180°) and west (270°) back to north (360° or 0°).

precession. The apparent slow movement of the celestial poles. This also means a shift of the celestial equator, and hence of the equinoxes. The vernal equinox moves by 50 seconds of arc per year, and is now in Pisces rather than Aries.

prime meridian. The meridian on the Earth's surface which passes through both poles and the Old Royal Observatory at Greenwich.

prominences. Masses of glowing gas, mainly hydrogen, rising from the Sun's surface. They were once (misleadingly) called Red Flames.

proper motion. The individual motion of a star on the celestial sphere.

proton. A fundamental particle with unit positive electrical charge.

200

Ptolemaic System. The old geocentric theory of the Solar System, with the Earth in the centre.

pulsar. A rapidly-rotating neutron star, sending out pulsed radio waves.

pulsating variable. A variable star in which the changes are intrinsic, so that the star swells and shrinks.

QSO (Quasi-Stellar Object). Alternative name for a quasar.

quadrant. An astronomical measuring instrument used in ancient times. It consisted of an arc graduated into 90°, with a sighting pointer.

quadrature. The position of the Moon or a planet when at right angles to the Sun as seen from the Earth. The Moon is at quadrature when at half-phase.

quantum. The energy possessed by one photon of light.

quasar. A very remote, super-luminous object; probably the core of a very active galaxy.

RR Lyrae variables. Variable stars with regular light curves and very short periods of a day or less. All are about 90 times as luminous as the Sun.

radial velocity. The towards-or-away movement of a celestial body: positive if the object is receding, negative if it is approaching.

radiant. The point in the sky from which the meteors of any particular shower appear to radiate.

radio telescope. An instrument used for collecting and analysing natural radio waves from space.

radio window. The region of the electromagnetic spectrum which is transparent to radio waves from space. It extends from about 20 metres down to a few millimetres.

radius vector. An imaginary line joining the centre of a planet (or comet) to the centre of the Sun. It also applies to satellites orbiting planets.

recurrent nova. A star which has been known to suffer more than one nova outburst.

red shift. The apparent increase in wavelength of the light of a body which is receding from the observer.

reflecting telescope. An optical telescope in which the light from the target object is collected by a curved mirror.

refraction. The 'bending' or change of direction of a ray of light when passing through a transparent surface. The shorter the wavelength, the greater the amount of refraction.

refractor. A telescope in which the light from the target object is collected by a lens (object-glass, or objective).

regolith. The outermost layer of the surface of the Moon or a planetary body.

resolving power. The ability of a telescope to separate objects which are close together. Generally speaking, the resolving power R of a telescope of aperture D is $R = 12/D$, where R is in seconds of arc and D in centimetres.

retardation. The difference in the time of moonrise on successive nights.

retrograde motion. Orbital or rotational movement in the sense opposite to that of the Earth's movement or rotation. A planet is said to move in an apparent retrograde direction when shifting from east to west on the celestial sphere.

reversing layer. The gaseous layer above the Sun's bright surface. It is responsible for the Fraunhofer lines in the solar spectrum.

right ascension (RA). The angular distance of a star from the vernal equinox, measured westward. It is usually given in units of time, and is the interval between the transit of the vernal equinox and the transit of the body concerned.

rill (otherwise spelled *rille*). Crack-line feature on the Moon's surface. True rills are collapse features, but some are really craterlet-chains. They are alternatively known as *clefts*.

satellites. Minor bodies orbiting planets.

Schmidt telescope. A type of telescope using a spherical mirror together with a special correcting plate. It can photograph a relatively wide area of the sky with a single exposure.

scintillation. The official name for 'twinkling' of a star or other celestial body.

secular acceleration. The apparent speeding-up of the Moon in its orbit, due to the gradual slowing down of the Earth's rotation by an average of 0.000 000 02 second per day.

Seyfert galaxies. Galaxies with small, bright nuclei and weak spiral arms. Most are active, and most are radio sources.

shooting-star. The popular name for a meteor.

sidereal period (or **periodic time**). The time taken for a body to complete one orbit round its primary: with the Earth, 365.25 days.

sidereal time. The local time reckoned according to the apparent rotation of the celestial sphere. The sidereal time is 0 hours when the vernal equinox crosses the observer's meridian.

solar time, apparent. The local time, reckoned according to the mean sun.

solar wind. A flow of charged particles streaming out from the Sun.

solstices. The times when the Sun is at its northernmost point in the sky (declination 23½°N), around 22 June, and at its southernmost point (declination 23½°S) around 22 December. The actual dates vary somewhat because of the complications in our calendar due to Leap Years.

spectroscope. An instrument for splitting up the light from a light-source, using a prism, diffraction grating or some equivalent device.

spectroscopic binary. A binary system whose components are too close together to be seen separately, but which can be detected spectroscopically because of the Doppler effect.

speculum. The main mirror of a reflecting telescope.

spring tide. The tide produced when the Sun and Moon are pulling in the same sense, i.e. at new and full moon.

steady-state theory. The theory that the universe has always existed, and will exist for ever, so that new material is being created spontaneously out of nothing. It has now been rejected.

style. The pointer or gnomon of a sundial.

summer triangle. An unofficial name for the pattern made by the stars Vega, Deneb and Altair. (Patrick Moore introduced it in a *Sky at Night* television programme many years ago, and everyone now seems to use it!)

superior planet. A planet moving round the Sun at a distance greater than that of the Earth (i.e. all the planets apart from Mercury and Venus).

supernova. A stellar explosion. (Type I) The complete destruction of the white dwarf component of a binary system. (Type II) The collapse of a very massive star.

synodic period. The interval between successive oppositions of a superior planet, or between successive inferior conjunctions of Mercury or Venus.

syzygy. The position of the Moon in its orbit when new or full.

tektites. Small, glassy objects found in a few restricted areas on the Earth. They may be special types of meteorites, but are more probably of terrestrial origin.

terminator. The boundary between the daylit and night hemispheres of the Moon or a planet.

transit. (a) The passage of a celestial object across the observer's meridian. (b) The projection of Mercury or Venus against the disk of the Sun.

transit instrument. A telescope mounted so that it can move only in declination; it is kept pointing at the meridian, and is used for timing the passages of stars across the meridian. The Airy transit instrument at Greenwich marks longitude 0°.

ultra-violet radiation. The region of the electro-magnetic spectrum between wavelength 10^{-8} metres and 4×10^{-7} metres (approximately 100 Ångströms to 4000 Ångströms). It lies between the visible and the X-ray range.

Van Allen zones. Zones round the Earth in which charged particles are trapped by the Earth's magnetic field. The outer zone is made up chiefly of electrons, the inner zone of protons.

variable stars. Stars which change in magnitude over relatively short periods.

vernal equinox. The First Point of Aries.

white dwarf. A small, very dense star which has exhausted its reserves of nuclear power.

Wilson effect. The foreshortening of a sunspot near the solar limb, so that the penumbra appears broadest toward the limb.

X-ray radiation. The electromagnetic radiations with wavelengths between about 10^{-12} metres and 10^{-8} metres (0.01 to 100 Ångströms), between the ultra-violet and gamma-ray regions.

year. The time taken for the Earth to complete one orbit round the Sun. The *sidereal year* (365.26 days) is the true revolution period of the Earth; the *tropical year* (365.24 days) is the interval between successive passages of the Sun across the vernal equinox.

zenith. The observer's overhead point (altitude 90°).

zenithal hourly rate. The number of naked-eye meteors from a particular shower which would be expected to be seen by an observer under ideal conditions, with the radiant of the shower at the zenith.

zenith distance. The angular distance of a celestial body from the zenith.

Zodiac. A belt stretching right round the sky, 8° to either side of the ecliptic, in which the Sun, Moon and planets (apart from Pluto) are always to be found.

Zodiacal light. A cone of light rising from the horizon, stretching along the ecliptic. It is due to the illumination of thinly-spread material in the main plane of the Solar System.

Zürich number (or **Wolf number**). A measure of sunspot activity. The formula is $Z = k(10g + n)$, where Z is the Zürich number, g is the number of groups, and n is the total number of individual spots; k is a constant, usually about 1, depending upon the experience and equipment of the observer.

Index

203